Android Application Security Essentials

Write secure Android applications using the most up-to-date techniques and concepts

Pragati Ogal Rai

PUBLISHING

BIRMINGHAM - MUMBAI

Android Application Security Essentials

First published: August 2013

Production Reference: 1140813

Published by Packt Publishing Ltd.
Livery Place
35 Livery Street
Birmingham B3 2PB, UK.

ISBN 978-1-84951-560-3

www.packtpub.com

Cover Image by Karl Moore (karl@karlmoore.co.uk)

Credits

Author

Pragati Ogal Rai

Reviewer

Alessandro Parisi

Acquisition Editor

Martin Bell

Lead Technical Editor

Madhuja Chaudhari

Technical Editors

Sampreshita Maheshwari

Larissa Pinto

Project Coordinator

Hardik Patel

Proofreader

Maria Gould

Indexer

Priya Subramani

Graphics

Abhinash Sahu

Ronak Druv

Production Coordinator

Prachali Bhiwandkar

Cover Work

Prachali Bhiwandkar

Foreword

When I first began working at GO Corporation in the early 1990s, the state of the art in mobile computing was an 8-lb, clipboard sized device with minimal battery life and an optional 9600 baud modem. But the vision that drove that device could just as easily be applied to the newest Android and iOS devices released this year: the desire for an integrated, task-centric computing platform with seamless connectivity. Back then, we thought that the height of that vision would be the ability to "send someone a fax from the beach." By the time I helped AOL deliver AIM, its instant messaging client, as one of the launch titles for Apple's iPhone App Store in 2008, that vision was already on its way to becoming a reality. But even at that time, just a few years ago, we couldn't have predicted what a tremendous effect these devices and the app ecosystem they spawned would have on our day-to-day lives.

Today, mobile devices are everywhere. They entertain us, they help us pass the time; and of course, they help us keep in touch (though perhaps not so much through fax). The Android operating system by Google is one of the driving forces behind this revolution, having been adopted by hundreds of device vendors and installed on nearly a billion devices worldwide. But as these mobile devices pervade every corner of our lives, keeping them — and their users — secure becomes critical. That's why this book is so important.

Viruses, Trojan horses, and malware may still be more prevalent on desktop platforms than they are on mobile. But the growth of the mobile market has meant a sharp rise in malicious software; anti-virus maker Kaspersky reports thousands of new programs detected each month. And today's smartphones and tablets represent an irresistible honey pot to the would-be attacker. Personal information, financial data, passwords, and social graphs, even up to the moment location data — everything that makes these devices so valuable to consumers is also what makes them such an attractive target to pranksters and data thieves. As developers, it's our responsibility to be good stewards of the information our users have entrusted to us. And the open and integrated nature of the Android operating system means it's much more important that each of us do our part to secure our applications and services.

Security can't be just a checkbox or an afterthought; it needs to be part of the design, and woven throughout the implementation of your application. I know Pragati Rai understands this intimately, having worked on this problem from both the perspective of the OS and the application developer. That's why she's so well positioned to write this book. She is able to look at the entirety of the Android ecosystem, from device to kernel to application, and present clear and actionable steps developers can take to secure their applications and data, along with source code that illustrates their use and methodologies to test their effectiveness. Moreover, she goes beyond the bits and bytes to explore security policy and best practices that can balance a developer's desire to use personal information with the user's desire to protect it.

The convergence of powerful mobile devices, ubiquitous social media, and the ability to transmit, store, and consume vast quantities of data has raised the stakes for everyone when it comes to mobile security. But security is like the air we breathe; we don't really think about it until it's gone, and by then it's often too late — too late to protect our users, and too late to protect the developer's reputation and business. So, it's critically important for every Android developer to understand the role they play in keeping users safe in this complex and ever-changing landscape.

As a developer and a user myself, I'm thankful that Pragati has taken the time to write such a comprehensive and informative guide to help us navigate this space, and I'm hopeful that her lessons will enable Android developers everywhere to give us the engaging and innovative applications we crave, while maintaining the security and trust we expect and deserve.

Edwin Aoki
Technology Fellow, PayPal

About the Author

Pragati Ogal Rai is a technologist with more than 14 years of experience in mobile operating systems, mobile security, mobile payments, and mobile commerce. From working as a platform security engineer with *Motorola Mobility*, to designing and developing PayPal's mobile offerings, she has an extensive end-to-end experience in all aspects of mobile technology.

Pragati has a dual Master's in Computer Science and has taught and trained computer science students at different levels. She is a recognized speaker at international technology events.

My sincere thanks to the entire Packt Publishing team for bringing this book to life. Special thanks to Hardik Patel, Madhuja Chaudhari, and Martin Bell for working diligently with me throughout the writing of this book and accommodating my crazy schedule. I want to acknowledge Alessandro Parisi for his candid comments and suggestions to improve the quality of the book.

Thanks to the thriving and vibrant community of Android developers who are the reason behind this book.

A big thank you to all my friends and family for encouraging me to write this book. In particular, I want to thank two families, the Khannas and the Kollis, who were my pillars of support during the writing of this book. Special thanks to Selina Garrison for her guidance and for being there for me. Last but most importantly, I want to thank my husband, Hariom Rai, and my son, Arnav Rai, who constantly encouraged, supported, and cheered me in their own ways as I wrote this book. Without them this book could not have been completed.

About the Reviewer

Alessandro Parisi is an enterprise software architect and an ethical hacker, working as an IT consultant for nearly 20 years now, keen on experimenting non-conventional solutions to problem solving in complex and dynamic contexts, mixing new technologies with lateral thinking and a holistic approach.

Founder of InformaticaSicura.com, specializing in IT security consultancy, he is the curator of Hacking Wisdom column appearing on the blog informaticasicura.altervista.org.

He is also the author of *Sicurezza Informatica e Tutela della Privacy*, published by Istituto Poligrafico e Zecca dello Stato, Italy, 2006.

I would like to acknowledge Ilaria Sinisi for her support and patience. Thank you very much, Ilaria.

www.PacktPub.com

Support files, eBooks, discount offers and more

You might want to visit www.PacktPub.com for support files and downloads related to your book.

Did you know that Packt offers eBook versions of every book published, with PDF and ePub files available? You can upgrade to the eBook version at www.PacktPub.com and as a print book customer, you are entitled to a discount on the eBook copy. Get in touch with us at service@packtpub.com for more details.

At www.PacktPub.com, you can also read a collection of free technical articles, sign up for a range of free newsletters and receive exclusive discounts and offers on Packt books and eBooks.

http://PacktLib.PacktPub.com

Do you need instant solutions to your IT questions? PacktLib is Packt's online digital book library. Here, you can access, read and search across Packt's entire library of books.

Why Subscribe?
- Fully searchable across every book published by Packt
- Copy and paste, print and bookmark content
- On demand and accessible via web browser

Free Access for Packt account holders

If you have an account with Packt at www.PacktPub.com, you can use this to access PacktLib today and view nine entirely free books. Simply use your login credentials for immediate access.

To my mom, Rekha Ogal. I love you, mom, and miss you very much.
May you rest in peace.

Table of Contents

Preface

In today's techno-savvy world, more and more of our lives are going digital and all this information is accessible anytime and anywhere using mobile devices. There are thousands of apps available for users to download and play with. With so much information easily accessible using application on the mobile devices, the biggest challenge is to secure the users' private information and respect their privacy.

The first Android phone came out in 2009. The mobile ecosystem has not been the same since then. The openness of the platform and a far less restrictive application model created excitement in the developer community and also fostered innovation and experimentation. But just as every coin has two sides, so does openness. The Android platform irked the imagination of the so-called bad guys. Android provides a perfect test bed for them to try out their ideas. It is thus of great importance not only as a developer, but also as a consumer, to be aware of Android's security model and how to use it judiciously to protect yourself and your consumers.

Android Application Security Essentials is a deep dive into Android security from the kernel level to the application level, with practical hands-on examples, illustrations, and everyday use cases. This book will show you how to secure your Android applications and data. It will equip you with tricks and tips that will come in handy as you develop your applications.

You will learn the overall security architecture of the Android stack. Securing components with permissions, defining security in manifest file, cryptographic algorithms, and protocols on Android stack, secure storage, security focused testing, and protecting enterprise data on device is also discussed in detail. You will also learn how to be security aware when integrating newer technologies and use cases such as NFC and mobile payments into your Android applications.

What this book covers

Chapter 1, Android Security Model – the Big Picture, focuses on the overall security of the Android stack all the way from platform security to application security. This chapter will form a baseline on which the subsequent chapters will be built upon.

Chapter 2, Application Building Blocks, introduces application components, permissions, manifest files, and application signing from a security perspective. These are all basic components of an Android application and knowledge of these components is important to build our security knowledge.

Chapter 3, Permissions, talks about existing permissions in the Android platform, how to define new permissions, how to secure application components with permissions, and provides an analysis of when to define a new permission.

Chapter 4, Defining the Application's Policy File, drills down into the mechanics of the manifest file, which is the application's policy file. We talk about tips and tricks to tighten up the policy file.

Chapter 5, Respect Your Users, covers best practices on handling users' data properly. This is important as a developer's reputation depends on user reviews and ratings. The developer should also be careful about handling user private information carefully so as not to fall into legal traps.

Chapter 6, Your Tools – Crypto APIs, discusses cryptographic capabilities provided by the Android platform. These include symmetric encryption, asymmetric encryption, hashing, cipher modes, and key management.

Chapter 7, Securing Application Data, is all about secure storage of application data both at rest and in transit. We talk about how private data is sandboxed with the application, how to securely store data on the device, on external memory cards, drives, and databases.

Chapter 8, Android in the Enterprise, talks about device security artifacts that are provided by the Android Platform and what they mean to an application developer. This chapter is of special interest to enterprise application developers.

Chapter 9, Testing for Security, focuses on designing and developing security-focused test cases.

Chapter 10, Looking into the Future, discusses upcoming use cases in the mobile space and how it affects Android, especially from a security perspective.

What you need for this book

This book is much more valuable if you have an Android environment set up and can play with the concepts and examples discussed in this book. Please refer to developer.android.com for detailed instructions on how to set up your environment and get started with Android development. If you are interested in kernel development, please refer to source.android.com.

At the time of writing this book, Jelly Bean (Android 4.2, API level 17) is the latest release. I have tested all my code snippets on this platform. Ever since the first release of Cupcake in 2009, Google has been continuously enhancing the security of Android with successive releases. For example, remote wipe and device management APIs were added in Android 2.2 (API level 8) to make Android more appealing to the business community. Whenever relevant, I have referenced the release that started supporting a particular feature.

Who this book is for

This book is an excellent resource for anyone interested in mobile security. Developers, test engineers, engineering managers, product managers, and architects may use this book as a reference when designing and writing their applications. Senior management and technologists may use this book to gain a broader perspective on mobile security. Some prior knowledge of development on the Android stack is desirable but not required.

Conventions

In this book, you will find a number of styles of text that distinguish between different kinds of information. Here are some examples of these styles, and an explanation of their meaning.

Code words in text are shown as follows: "The PackageManager class handles the task of installing and uninstalling the application."

A block of code is set as follows:

```
<intent-filter>
  <action android:name="android.intent.action.MAIN" />
  <category android:name="android.intent.category.LAUNCHER" />
</intent-filter>
```

When we wish to draw your attention to a particular part of a code block, the relevant lines or items are set in bold:

```
Intent intent = new Intent("my-local-broadcast");

Intent.putExtra("message", "Hello World!");

LocalBroadcastManager.getInstance(this).sendBroadcast(intent);
```

Any command-line input or output is written as follows:

```
dexdump -d -f -h data@app@com.example.example1-1.apk@classes .dex > dump
```

New terms and **important words** are shown in bold. Words that you see on the screen, in menus or dialog boxes for example, appear in the text like this: "clicking the **Next** button moves you to the next screen".

 Warnings or important notes appear in a box like this.

 Tips and tricks appear like this.

Reader feedback

Feedback from our readers is always welcome. Let us know what you think about this book—what you liked or may have disliked. Reader feedback is important for us to develop titles that you really get the most out of.

To send us general feedback, simply send an e-mail to feedback@packtpub.com, and mention the book title via the subject of your message.

If there is a topic that you have expertise in and you are interested in either writing or contributing to a book, see our author guide on www.packtpub.com/authors.

Customer support

Now that you are the proud owner of a Packt book, we have a number of things to help you to get the most from your purchase.

Errata

Although we have taken every care to ensure the accuracy of our content, mistakes do happen. If you find a mistake in one of our books—maybe a mistake in the text or the code—we would be grateful if you would report this to us. By doing so, you can save other readers from frustration and help us improve subsequent versions of this book. If you find any errata, please report them by visiting `http://www.packtpub.com/submit-errata`, selecting your book, clicking on the **errata submission form** link, and entering the details of your errata. Once your errata are verified, your submission will be accepted and the errata will be uploaded on our website, or added to any list of existing errata, under the Errata section of that title. Any existing errata can be viewed by selecting your title from `http://www.packtpub.com/support`.

Piracy

Piracy of copyright material on the Internet is an ongoing problem across all media. At Packt, we take the protection of our copyright and licenses very seriously. If you come across any illegal copies of our works, in any form, on the Internet, please provide us with the location address or website name immediately so that we can pursue a remedy.

Please contact us at `copyright@packtpub.com` with a link to the suspected pirated material.

We appreciate your help in protecting our authors, and our ability to bring you valuable content.

Questions

You can contact us at `questions@packtpub.com` if you are having a problem with any aspect of the book, and we will do our best to address it.

1
The Android Security Model – the Big Picture

Welcome to the first chapter of *Android Application Security Essentials*!

The Android stack is different in many ways. It is open; more advanced than some of the other platforms, and imbibes the learning from attempts to develop a mobile platform in the past. In this first chapter, we introduce the basics of the Android security model from the kernel all the way to the application level. Each security artifact introduced in this chapter is discussed in greater detail in the following chapters.

We kick off the chapter with explaining why install time application permission evaluation is integral to the security of the Android platform and user data. Android has a layered architecture and a security evaluation of each architectural layer is discussed in this chapter. We end the chapter with a discussion of core security artifacts such as application signing, secure data storage on the device, crypto APIs, and administration of an Android device.

Installing with care

One of the differentiating factors of Android from other mobile operating systems is the install time review of an application's permissions. All permissions that an application requires have to be declared in the application's manifest file. These permissions are capabilities that an application requires for functioning properly. Examples include accessing the user's contact list, sending SMSs from the phone, making a phone call, and accessing the Internet. Refer *Chapter 3*, *Permissions*, for a detailed description of the permissions.

When a user installs an application, all permissions declared in the manifest file are presented to the user. A user then has the option to review the permissions and make an informed decision to install or not to install an application. Users should review these permissions very carefully as this is the only time that a user is asked for permissions. After this step, the user has no control on the application. The best a user can do is to uninstall the application. Refer to the following screenshot for reference. In this example, the application will track or access the user location, it will use the network, read the user's contact list, read the phone state, and will use some development capabilities. When screening this application for security, the user must evaluate if granting a certain power to this application is required or not. If this is a gaming application, it might not need development tool capabilities. If this is an educational application for kids, it should not need access to the contact list or need to access the user location. Also be mindful of the fact that a developer can add their own permissions especially if they want to communicate with other applications that they have developed as well and may be installed on the device. It is the onus of the developer to provide a clear description of such permissions.

At install time, the framework ensures that all permissions used in the application are declared in the manifest file. The OS at runtime then enforces these permissions.

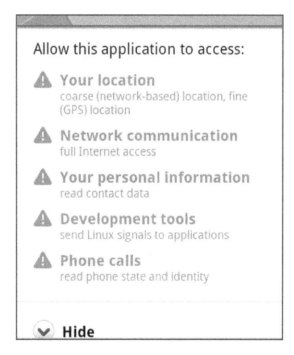

Android platform architecture

Android is a modern operating system with a layered software stack. The following figure illustrates the layers in Android's software stack. This software stack runs on top of the device hardware. Android's software stack can run on many different hardware configurations such as smartphones, tablets, televisions, and even embedded devices such as microwaves, refrigerators, watches, and pens. Security is provided at every layer, creating a secure environment for mobile applications to live and execute. In this section, we will discuss the security provided by each layer of the Android stack.

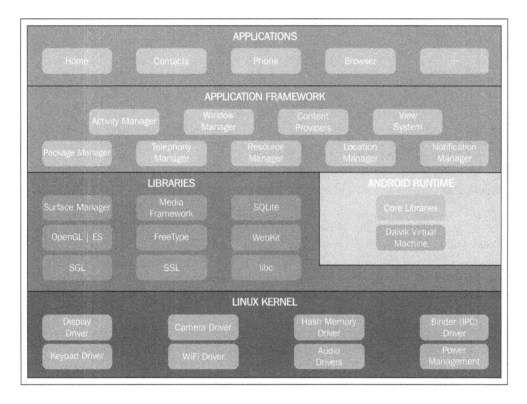

Linux kernel

On top of the device hardware sits the **Linux kernel**. The Linux kernel has been in use for decades as a secure multi-user operating system, isolating one user from the other. Android uses this property of Linux as the basis of Android security. Imagine Android as a multi-user platform where each user is an application and each application is isolated from each other. The Linux kernel hosts the device drivers such as drivers for bluetooth, camera, Wi-Fi, and flash memory. The kernel also provides a mechanism for secure **Remote Procedure Calls** (RPC).

As each application is installed on the device, it is given a unique **User Identification (UID)** and **Group Identification (GID)**. This UID is the identity of the application for as long as it is installed on the device.

Refer to the following screenshot. In the first column are all the application UIDs. Notice the highlighted application. Application com.paypal.com has the UID app_8 and com.skype.com has the UID app_64. In the Linux kernel, both these applications run in their own processes with this ID.

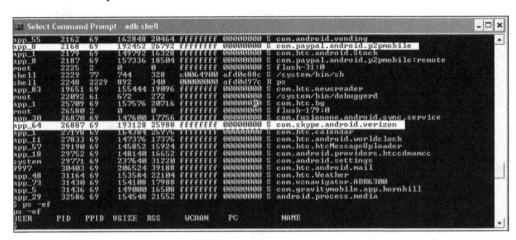

Refer to the next screenshot. When we give the id command in the shell, the kernel displays the UID, GID, and the groups the shell is associated with. This is the process sandbox model that Android uses to isolate one process from the other. Two processes can share data with each other. The proper mechanics to do so are discussed in *Chapter 4, Defining the Application's Policy File*.

Although most Android applications are written in Java, it is sometimes required to write native applications. Native applications are more complex as developers need to manage memory and device-specific issues. Developers can use the Android NDK toolset to develop parts of their application in C/C++. All native applications conform to Linux process sandboxing; there is no difference in the security of a native application and Java application. Bear in mind that just as with any Java application, proper security artifacts such as encryption, hashing, and secure communication are required.

Middleware

On top of the Linux kernel sits the middleware that provides libraries for code execution. Examples of such libraries are `libSSL`, `libc`, `OpenGL`. This layer also provides the runtime environment for Java applications.

Since most users write their apps on Android in Java, the obvious question is: does Android provide a **Java virtual machine**? The answer to this question is no, Android does not provide a Java virtual machine. So a **Java Archive (JAR)** file will not execute on Android, as Android does not execute byte code. What Android does provide is a **Dalvik virtual machine**. Android uses a tool called `dx` to convert byte codes to **Dalvik Executable (DEX)**.

Dalvik virtual machine

Originally developed by Dan Bornstein, who named it after the fishing village of Dalvik in Iceland where some of his ancestors lived, Dalvik is a register-based, highly optimized, open-sourced virtual machine. Dalvik does not align with Java SE or Java ME and its library is based on **Apache Harmony**.

Each Java application runs in its own VM. When the device boots up, a nascent process called **Zygote** spawns a VM process. This Zygote then forks to create new VMs for processes on request.

The main motivation behind Dalvik is to reduce memory footprint by increased sharing. The constant pool in Dalvik is thus a shared pool. It also shares core, read only libraries between different VM processes.

Dalvik relies on the Linux platform for all underlying functionality such as threading and memory management. Dalvik does have separate garbage collectors for each VM but takes care of processes that share resources.

Dan Bornstein made a great presentation about Dalvik at Google IO 2008. You can find it at `http://www.youtube.com/watch?v=ptjedOZEXPM`. Check it out!

Application layer

Application developers developing Java-based applications interact with the application layer of the Android stack. Unless you are creating a native application, this layer will provide you with all the resources to create your application.

We can further divide this application layer into the application framework layer and the application layer. The application framework layer provides the classes that are exposed by the Android stack for use by an application. Examples include the Activity manager that manages the life-cycle of an Activity, the package manager that manages the installing and uninstalling of an application, and the notification manager to send out notifications to the user.

The application layer is the layer where applications reside. These could be system applications or user applications. System applications are the ones that come bundled with the device such as mail, calendar, contacts, and browser. Users cannot uninstall these applications. User applications are the third party applications that users install on their device. Users can install and uninstall these applications at their free will.

Android application structure

To understand the security at the application layer, it is important to understand the Android application structure. Each Android application is created as a stack of components. The beauty of this application structure is that each component is a self-contained entity in itself and can be called exclusively even by other applications. This kind of application structure encourages the sharing of components. The following figure shows the anatomy of an Android application that consists of activities, services, broadcast receivers, and content providers:

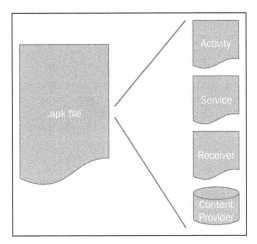

Android supports four kinds of components:

- **Activity**: This component is usually the UI part of your application. This is the component that interacts with the user. An example of the Activity component is the login page where the user enters the username and password to authenticate against the server.

- **Service**: This component takes care of the processes that run in the background. The Service component does not have a UI. An example could be a component that synchronizes with the music player and plays songs that the user has pre-selected.

- **Broadcast Receiver**: This component is the mailbox for receiving messages from the Android system or other applications. As an example, the Android system fires an Intent called BOOT_COMPLETED after it boots up. Application components can register to listen to this broadcast in the manifest file.

- **Content Provider**: This component is the data store for the application. The application can also share this data with other components of the Android system. An example use case of the Content Provider component is an app that stores a list of items that the user has saved in their wish list for shopping.

All the preceding components are declared in the AndroidManifest.xml (manifest) file. In addition to the components, the manifest file also lists other application requirements such as the minimum API level of Android required, user permissions required by the application such as access to the Internet and reading of the contact list, permission to use hardware by the application such as Bluetooth and the camera, and libraries that the application links to, such as the Google Maps API. *Chapter 4, Defining the Application's Policy File*, discusses the manifest file in greater detail.

Activities, services, content providers, and broadcast receivers all talk to each other using Intents. Intent is Android's mechanism for asynchronous **inter-process communication** (**IPC**). Components fire off Intent to do an action and the receiving component acts upon it. There are separate mechanisms for delivering Intents to each type of components so the Activity Intents are only delivered to activities and the broadcast Intents are only delivered to broadcast receivers. Intent also includes a bundle of information called the Intent object that the receiving component uses to take appropriate action. It is important to understand that Intents are not secure. Any snooping application can sniff the Intent, so don't put any sensitive information in there! And imagine the scenario where the Intent is not only sniffed but also altered by the malicious application.

As an example, the following figure shows two applications, **Application A** and **Application B**, both with their own stack of components. These components can communicate with each other as long as they have permissions to do so. An **Activity** component in **Application A** can start an **Activity** component in **Application B** using startActivity() and it can also start its own **Service** using startService().

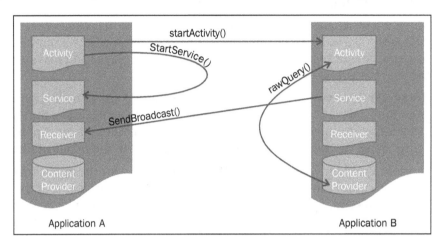

At the application level, Android components follow the permission-based model. This means that a component has to have appropriate permission to call the other components. Although Android provides most of the permissions that an application might need, developers have the ability to extend this model. But this case should be rarely used.

Additional resources such as bitmaps, UI layouts, strings, and so on, are maintained independently in a different directory. For the best user experience, these resources should be localized for different locales, and customized for different device configurations.

The next three chapters talk about the application structure, the manifest file, and the permission model in detail.

Application signing

One of the differentiating factors of Android is the way Android applications are signed. All applications in Android are self-signed. There is no requirement to sign the applications using a certificate authority. This is different from traditional application signing where a signature identifies the author and bases trust upon the signature.

The signature of the application associates the app with the author. If a user installs multiple applications written by the same author and these applications want to share each other's data, they need to be associated with the same signature and should have a SHARED_ID flag set in the manifest file.

The application signature is also used during the application upgrade. An application upgrade requires that both applications have the same signature and that there is no permission escalation. This is another mechanism in Android that ensures the security of applications.

As an application developer, it is important to keep the private key used to sign the application secure. As an application author, your reputation depends on it.

Data storage on the device

Android provides different solutions for secure data storage on devices. Based on the data type and application use case, developers can choose the solution that fits best.

For primitive data types such as ints, booleans, longs, floats, and strings, which need to persist across user sessions, it is best to use shared data types. Data in shared preferences is stored as a key-value pair that allows developers to save, retrieve, and persist data.

All application data is stored along with the application in the sandbox. This means that this data can be accessed only by that application or other applications with the same signature that have been granted the right to share data. It is best to store private data files in this memory. These files will be deleted when the application is uninstalled.

For large datasets, developers have an option to use the SQLite database that comes bundled with the Android software stack.

All Android devices allow users to mount external storage devices such as SD cards. Developers can write their application such that large files can be stored on these external devices. Most of these external storage devices have a VFAT filesystem, and Linux access control does not work here. Sensitive data should be encrypted before storing on these external devices.

Starting with Android 2.2 (API 8), APKs can be stored on external devices. Using a randomly generated key, the APK is stored within an encrypted container called the `asec` file. This key is stored on the device. The external devices on Android are mounted with `noexec`. All DEX files, private data, and native shared libraries still reside in the internal memory.

Wherever network connection is possible, developers can store data on their own web servers as well. It is advisable to store data that can compromise the user's privacy on your own servers. An example of such an application is banking applications where user account information and transaction details should be stored on a server rather than user's devices.

Chapter 7, *Securing Application Data*, discusses the data storage options on Android devices in great detail.

Rights protected content such as video, e-books, and music, can be protected on Android using the DRM framework API. Application developers can use this DRM framework API to register the device with a DRM scheme, acquire licenses associated with content, extract constraints, and associate relevant content with its license.

Crypto APIs

Android boasts of a comprehensive crypto API suite that application developers can use to secure data, both at rest and in transit.

Android provides APIs for symmetric and asymmetric encryption of data, random number generation, hashing, message authentication codes, and different cipher modes. Algorithms supported include DH, DES, Triple DES, RC2, and RC5.

Secure communication protocols such as SSL and TLS, in conjunction with the encryption APIs, can be used to secure data in transit. Key management APIs including the management of X.509 certificates are provided as well.

A system key store has been in use since Android 1.6 for use by VPN. With Android 4.0, a new API called `KeyChain` provides applications with access to credentials stored there. This API also enables the installation of credentials from X.509 certificates and PKCS#12 key stores. Once the application is given access to a certificate, it can access the private key associated with the certificate.

Crypto APIs are discussed in detail in *Chapter 6*, *Your Tools – Crypto APIs*.

Device Administration

With the increased proliferation of mobile devices in the workplace, Android 2.2 introduced the **Device Administration API** that lets users and IT professionals manage devices that access enterprise data. Using this API, IT professionals can impose system level security policies on devices such as remote wipe, password enablement, and password specifics. Android 3.0 and Android 4.0 further enhanced this API with polices for password expiration, password restrictions, device encryption requirement, and to disable the camera. If you have an email client and you use it to access company email on your Android phone, you are most probably using the Device Administration API.

The Device Administration API works by enforcing security policies. The `DevicePolicyManager` lists out all the policies that a Device Administrator can enforce on the device.

A Device Administrator writes an application that users install on their device. Once installed, users need to activate the policy in order to enforce the security policy on the device. If the user does not install the app, the security policy does not apply but the user cannot access any of the features provided by the app. If there are multiple Device Administration applications on the device, the strictest policy prevails. If the user uninstalls the app, the policy is deactivated. The application may decide to reset the phone to factory settings or delete data based on the permissions it has as it uninstalls.

We will discuss Device Administration in greater detail in *Chapter 8*, *Android in the Enterprise*.

Summary

Android is a modern operating system where security is built in the platform. As we learned in this chapter, the Linux kernel, with its process isolation, provides the basis of Android's security model. Each application, along with its application data, is isolated from other processes. At the application level, components talk to each other using Intents and need to have appropriate privileges to call other components. These permissions are enforced in the Linux kernel that has stood the test of time as a secure multiuser operating system. Developers have a comprehensive set of crypto APIs that secure user data.

With this basic knowledge of the Android platform, let's march to the next chapter and understand application components and inter-component communication from a security standpoint. Good luck!

2
Application Building Blocks

This chapter focuses on the building blocks of an Android application, namely, the application components and the inter-component communication. There are four types of components in the Android system: Activities, Services, Broadcast Receivers, and Content Providers. Each component is specially designed to accomplish a specific task. A collection of these components makes an Android application. These components talk to each other using Intents which is Android's mechanism for inter-process communication.

There are several books that discuss how to build Android components and Intents. In fact, the Android developer website does a pretty good job introducing programming using these components as well. So in this chapter, instead of covering the implementation details, our objective is to discuss the security aspects of each component and how to define and use component and Intents securely in an application to protect our reputation as a developer and the privacy of our consumers.

Components and Intents are the focus of this chapter. For each Android component, we will cover component declaration, permissions associated with the component, and other security considerations specific to that particular component. We will discuss different types of Intents and the best Intent to use in a particular context.

Application components

As we have briefly touched in *Chapter 1, Android Security Model – the Big Picture*, an Android application is a loosely bound stack of application components. Application components, manifest file, and application resources are packaged in an **Application Package Format** .apk file. An **APK** file is essentially a ZIP file formatted in JAR file format. The Android system only recognizes the APK format, so all packages have to be in the APK format to be installed on the Android device. An APK file is then signed with the developer's signature to assert the authorship. The `PackageManager` class handles the task of installing and uninstalling the application.

In this section, we will talk about the security of each of the components in detail. This includes the declaration of a component in the manifest file, so we prune loose ends and other security considerations that are unique to each component.

Activity

An Activity is the application component that usually interacts with the user. An Activity extends the `Activity` class and is implemented as views and fragments. Fragments were introduced in **Honeycomb** to address the issue of different screen sizes. On a smaller screen, a fragment is shown as a single Activity and allows the user to navigate to the second Activity to display the second fragment. Fragments and threads spun by an Activity run in the context of the Activity. So if the Activity is destroyed, the fragments and threads associated with it will be destroyed as well.

An application can have several activities. It is best to use an Activity to focus on a single task and to create different activities for individual tasks. For example, if we are creating an application that lets users order books on a website, it is best to create an Activity to log the user in, another Activity for searching books in the database, another Activity for entering ordering information, another one for entering payment information, and so on. This style encourages Activity reuse within the application and by other applications installed on the device. The reuse of components has two major benefits. First, it helps to reduce bugs, as there is less duplication of code. Second, it makes the application more secure as there is less sharing of data between different components.

Activity declaration

Any Activity that an application uses has to be declared in the `AndroidManifest.xml` file. The following code snippet shows a login Activity and an order Activity declared in the manifest file:

```
<activity android:label="@string/app_name" android:name=".
LoginActivity">
  <intent-filter>
    <action android:name="android.intent.action.MAIN" />
    <category android:name="android.intent.category.LAUNCHER" />
  </intent-filter>
</activity>
<activity android:name=".OrderActivity" android:permission="com.
example.project.ORDER_BOOK" android:exported="false"/>
```

Note that `LoginActivity` is declared as a public Activity that may be launched by any other Activity in the system. The `OrderActivity` is declared as a private Activity (an Activity with no Intent filters is a private Activity to be invoked only by specifying its exact filename) that is not exposed outside the application. An additional `android:exported` tag can be used to specify if it is visible outside the application. A value of `true` makes the Activity visible outside the application, and a value of `false` does otherwise. The Intent Filter is discussed later in this chapter.

All the Activities can be secured by permissions. In the preceding example, the `OrderActivity`, besides being private, is also protected by a permission `com.example.project.ORDER_BOOK`. Any component that tries to invoke `OrderActivity` should have this custom permission to invoke it.

Usually, whenever an Activity is launched, it runs in the process of the application that declared it. Setting the `android:multiprocess` attribute to `true` lets an Activity run in a process different from the application. These process specifics can be defined using the `android:process` attribute. If the value of this attribute starts with a colon (`:`), a new process private to the application is created; if it starts with a lowercase character, the Activity runs in a global process.

The `android:configChanges` tag lets the application handle Activity restarts due to listed configuration changes. Such changes include changes in locale, plugging an external keyboard, and SIM changes.

Saving the Activity state

All the Activities are managed by the system in the **activity stack**. The Activity currently interacting with the user runs in the foreground. The current Activity can then launch other Activity. Any Activity that is in the background may be killed by the Android system due to resource constraints. An Activity may also be restarted during configuration changes such as change in orientation from vertical to horizontal. As mentioned in the preceding section, an Activity can use the `android:configChanges` tag to handle some of these events itself. It is not encouraged as it may lead to inconsistencies.

The state of the Activity should be preserved before a restart happens. The lifecycle of an Activity is defined by the following methods:

```
public class Activity extends ApplicationContext {
    protected void onCreate(Bundle savedInstanceState);
    protected void onStart();
    protected void onRestart();
    protected void onResume();
```

```
    protected void onPause();
    protected void onStop();
    protected void onDestroy();
}
```

An Activity may override `onSaveInstanceState(Bundle savedInstanceState)` and `onRestoreInstanceState(Bundle savedInstanceState)`, to save and restore the instance values such as user preferences and unsaved text. The Android developer website, `http://www.developer.android.com`, illustrates this process beautifully with the following flowchart:

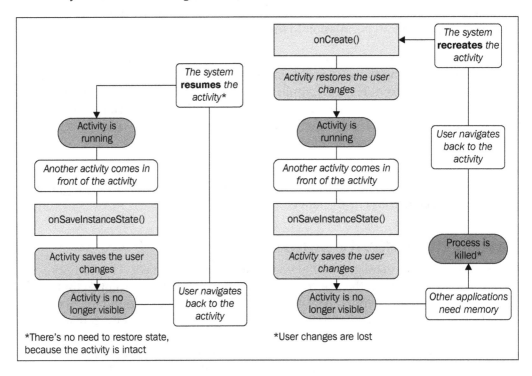

The following code snippet shows how an Activity may store and retrieve the preferred language, number of search results, and author name. User preferences are stored as a **Bundle**, which stores name-value pairs, when the Activity is killed. When the Activity restarts, this Bundle is passed to the `onCreate` method, which restores the Activity state. It is important to note that this method of storage does not persist application restarts.

```
@Override
public void onSaveInstanceState(Bundle savedInstanceState) {
    super.onSaveInstanceState(savedInstanceState);
```

```
    savedInstanceState.putInt("ResultsNum", 10);
    savedInstanceState.putString("MyLanguage", "English");
    savedInstanceState.putString("MyAuthor", "Thomas Hardy");
}

@Override
public void onRestoreInstanceState(Bundle savedInstanceState) {
    super.onRestoreInstanceState(savedInstanceState);
    int ResultsNum = savedInstanceState.getInt("ResultsNum");
    String MyLanguage = savedInstanceState.getString("MyLanguage");
    String MyAuthor = savedInstanceState.getString("MyAuthor");
}
```

Saving user data

As we discussed earlier, the Activities interact with the users so they may collect some user data. The data could be private to the application or shared with others. An example of such data could be the user's preferred language or book category. This kind of data is generally retained by the application to enhance the user experience. It is useful within the application itself and is not shared with other applications.

An example of shared data could be the wish list of books that the user keeps adding to the collection as users browse through the store. This data may or may not be shared with other applications.

Based on the privacy and kind of data, a different storage mechanism can be employed. An application can decide to use SharedPreferences, Content Provider, a file stored on internal or external memory, or even on the developer's own website to store this type of data. Content providers are discussed in this chapter. Other persistent data storage mechanisms are discussed in detail in *Chapter 7, Securing Application Data.*

Service

Unlike Activities, Services lack a visual interface and are used in the background for long running tasks. Ideally, a Service should keep running in the background even when the Activity that started it no longer exists. When the task is completed, a Service should stop by itself. Examples of tasks best suited for a Service are syncing with database, uploading or downloading files from the network, interacting with the music player to play tracks selected by the user, and global services that applications can bind to for information.

Securing a Service starts with the Service declaration in the manifest file. Next it is important to identify the correct Service for a use case and manage the lifecycle of a Service. This includes starting and stopping a Service and creating a worker thread to avoid blocking the application. In the next few sections, we will walk through each of these aspects. The last section of the chapter is about binders, which is the backbone for most of Android's IPC and enables the Service to be used in a client-server fashion.

Service declaration

All the Services that an application plans to start need to be declared in the manifest file. The Service declaration defines how a Service, once created, will run. The syntax of the `<service>` tag in the manifest file is shown in the following code snippet:

```
<service android:enabled=["true" | "false"]
        android:exported=["true" | "false"]
        android:icon="drawable resource"
        android:isolatedProcess=["true" | "false"]
        android:label="string resource"
        android:name="string"
        android:permission="string"
        android:process="string" >
    . . . . .
</service>
```

Based on the preceding declaration syntax, a Service that is private to the application, and runs in its global process to store books in the database, can be declared as follows:

```
<service
  android:name="bookService"
  android:process=":my_process"
  android:icon="@drawable/icon"
  android:label="@string/service_name" >
</service>
```

By default, a Service runs in the global process of the application. In case an application wants to start a Service in a different process, it may do so using the attribute `android:process`. If the value of this attribute starts with a colon (`:`), the Service starts in a new private process within the application. If the value starts with a lowercase, a new global process is created that is visible and accessible to all applications of the Android system. In the preceding example, the Service runs in its own global process. The application should have permissions to create such a process.

This `android:enabled` attribute defines if the Service can be instantiated by the system or not. The default value is `true`.

The `android:exported` attribute limits the exposure of the Service. A value of `true` means that this Service is visible outside the application. If the Service contains an Intent Filter then the Service is visible to other applications. The default value of this attribute is `true`.

To run the Service in an isolated process, devoid of all permissions, set the `android:isolatedProcess` attribute to `true`. In this case, the only way to interact with the Service is through binding to the Service. The default value of this attribute is `false`.

As with Activities, Services can be protected by permissions. These services are declared in the manifest file using the `android:permission` attribute. The invoking components need to have proper permission to invoke the Service, otherwise a `SecurityException` is thrown from the call.

Service modes

A Service can be used in two contexts. In the first case, a Service acts as a helper Service that a component can start to run long running tasks. Such a Service is called a **started service**. The second use case for a Service is as a provider of information to components of one or many applications. In this case, the Service runs in the background and the application components bind to the Service by calling `bindService ()`. Such a Service is called a **bound service**.

A started service extends either the `Service` class or the `IntentService` class. The main difference between the two approaches is the handling of multiple requests. When extending the `Service` class, the application needs to take care of handling multiple requests. This is done in the `onStartCommand()` method.

The `IntentService()` class makes it easier by queuing all the requests and processing them one at a time, so the developer does not need to take care of threading. If suitable for a use case, it is always better to use the `IntentService` class to avoid multithreading bugs. The `IntentService` class starts a worker thread for the task and requests are queued automatically. The task is done in `onHandleIntent` and that's it! The following is an example of an `IntentService` class:

```
public class MyIntentService extends IntentService {
  public MyIntentService() {
    super("MyIntentService");
  }
  @Override
```

```
    protected void onHandleIntent(Intent intent) {
      // TODO Auto-generated method stub
    }
  }
```

A bound service is the client server case where a Service acts as the server and clients bind to it for information. This is done using the `bindService()` method. When the clients are satisfied, they unbind themselves from the Service using `unbindService()`.

A bound service can cater to components of one application or components of different applications. A bound service that only caters to one application component can extend the `Binder` class and implements the `onBind()` method which returns the `IBinder` object. If a Service caters to multiple applications, a messenger or **Android Interface Definition Language (AIDL)** tool can be used to generate interfaces published by a Service. Using a messenger is easier to implement as it takes care of multithreading.

When binding to a Service, it is important to check the identity of the Service that the Activity is binding to. This can be done by explicitly specifying the Service name. If the Service name is not available, the client can check the identity of the Service it is connected to using `ServiceConnection.onServiceConnected()`. Another method is to use permission checks.

 For a started service the `onBind()` method returns null.

Lifecycle management

A Service can be started by any component using the `startService()` method and passing an Intent object as follows:

```
Intent intent = new Intent(this, MyService.class);
startService(intent);
```

Just like any other component, a started service can also be destroyed by the Android system to gather resources for the process that the user is interacting with. In such a scenario, the Service will be restarted based on the return value set in the `onStartCommand` method. The following is an example:

```
@Override
public int onStartCommand(Intent intent, int flags, int startId) {
  handleCommand(intent);
  // Let the service run until it is explicitly stopped
  return START_STICKY;
}
```

There are three options for restarting a Service:

- START_NOT_STICKY: This option indicates the Android system not to restart the Service unless there are pending Intents. Pending Intents are discussed later in this chapter. This option is best for cases where an unfinished job can be safely restarted and finished later.

- START_STICKY: This option indicates that a Service should be started by the system. If the initial Intent is lost, the onStartCommand() method is started with a null Intent. This is best for cases, where even if the initial Intent is lost, the Service can resume its task. An example is the music player that starts again once it is killed by the system.

- START_REDELIVER_INTENT: In this option, the Service is restarted and the pending Intent is redelivered to the Service onStartCommand(). An example is downloading a file over the network.

It is important to note that a Service is different from creating a thread. A thread is killed immediately when the component that spun it is killed. A Service by default runs in the global application thread and remains alive even if the invoking component is destroyed. If the Service is doing some time consuming activity such as downloading a huge file, it is prudent to do it in a separate thread to avoid blocking the application.

A started service runs in the application thread by default. Any blocking Activities should be done in a separate thread to avoid potential bottlenecks when running your application. The IntentService class takes care of this scenario by spawning a worker thread.

Both kinds of started services should stop themselves by calling stopSelf() when the task has completed. Any component can stop the Service as well by using the method stopService().

A bound service is destroyed by the system when no more clients are binding to it.

 A Service can be both started and bound. In this case, do not forget to call stopSelf() or stopService() to stop a Service from continuing to run in the background.

Binder

Binder is the backbone of most of Android's IPC. It is a kernel driver and all calls to Binder go through the kernel. The messenger is based on Binder as well. Binders can be confusing to implement and should only be used if the Service caters to multiple applications running in different processes and wants to handle multithreading itself. The Binder framework is integrated in the OS, so a process that intends to use a Service of another process needs to marshal the objects into primitives. The OS then delivers it across the process boundary. To make this task easier for developers, Android provides the AIDL. The following figure illustrates how Binder is the core of all Android IPC. A Binder is exposed by AIDL. Intents are implemented as Binders as well. But these intricacies are hidden from the user. As we move to bigger concentric circles, the implementation becomes more abstract.

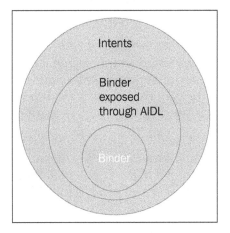

To create a bounded service using AIDL, we first create the AIDL file. Then, using the Android SDK tools, we generate the interface. This interface contains the `stub` method that extends the `android.os.Binder` class and implements the `onTransact()` method. The client receives a reference to the Binder interface and calls its `transact()` method. Data flows through this channel as a `Parcel` object. A `Parcel` object is serializable so it can effectively cross process boundaries.

 The `Parcel` objects are defined for high performance IPC transport, so they should not be used for general-purpose serialization.

If multiple processes are using the Service, beware not to change your AIDL once you have exposed it, as other applications might be using it as well. If this change is absolutely necessary then it should at least be backward compatible.

Binders are globally unique in the system and references to binders can be used as a shared secret to verify a trusted component. It is always a good idea to keep Binders private. Anyone who has a reference to the Binder can make calls to it and can call the `transact()` method. It is up to the Service to respond to the request. For example, Zygote, the system Service, exposes a Binder that any Activity can bind to. But calling its `transact()` method does not mean it will be entertained.

Binder can run in the same process or different process based on the `android:process` attribute of the `<service>` tag.

A Binder provides the identity of the calling component and its permission securely through the kernel. The identity of the caller can be checked using the methods `getCallingPid()` and `getCallingUid()` of the Binder. A Binder in turn can call other Binders which in this case can use the identity of the calling Binder. To check the permission of the caller, `Context.checkCallingPermission()` can be used. To check if the caller or Binder itself has a particular permission, `Context.checkCallingOrSelfPermission()` can be used.

Content Provider

Android system uses Content Providers for data storage such as contact list, calendar, and word dictionary. A Content Provider is Android's mechanism to handle structured data across process boundaries. It can be used within an application as well.

In most cases, the Content Provider's data is stored in the SQL database. The identifier `_id` is used as the primary key. As with SQL, users access data by writing queries. These can be `rawQuery()` or `query()` depending on whether they are raw SQL statements or structured queries. The return type of a query is a `Cursor` object that points to one of the rows of the results. Users can use helper methods such as `getCount()`, `moveToFirst()`, `isAfterLast()`, and `moveToNext()` to navigate multiple rows. `Cursor` needs to be closed using `close()` once the task is completed.

Providers support many different types of data including integer, long, float, double, and BLOB (Binary Large Object) implemented as a 64 KB array. Providers can also return standard or MIME types. An example of a standard MIME type is `text/html`. For custom MIME types, the value is always `vnd.android.cursor.dir` and `vnd.android.cursor.item` for multiple and single rows respectively.

The following figure illustrates a Content Provider that can abstract a database, a file, or even a remote server. Other components of the application can access it. So can other application components, provided they have appropriate permissions.

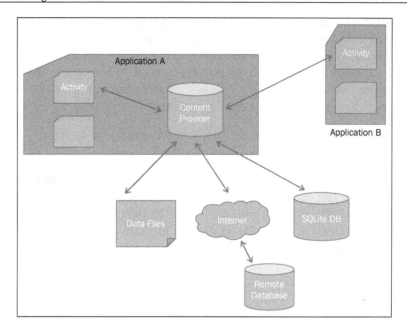

The following sections discuss the proper declaration of a provider, defining appropriate permissions, and avoiding common security pitfalls that are necessary for the secure access of provider data.

Provider declaration

Any provider that the application wants to use has to be declared in the manifest file. The syntax of the provider tag is as follows:

```
<provider android:authorities="list"
          android:enabled=["true" | "false"]
          android:exported=["true" | "false"]
          android:grantUriPermissions=["true" | "false"]
          android:icon="drawable resource"
          android:initOrder="integer"
          android:label="string resource"
          android:multiprocess=["true" | "false"]
          android:name="string"
          android:permission="string"
          android:process="string"
          android:readPermission="string"
          android:syncable=["true" | "false"]
          android:writePermission="string" >
    . . . . . . .
</provider>
```

Based on the preceding declaration syntax, a custom provider that maintains a list of books in the user's wish list can be declared as follows. The provider has read and write permissions and the client can request for temporary access to the path /figures.

```
<provider
    android:authorities="com.example.android.books.contentprovider"
    android:name=".contentprovider.MyBooksdoContentProvider"
    android:readPermission="com.example.android.books.DB_READ"
    android:writePermission="com.example.android.book.DB_WRITE">

    <grant-uri-permission android:path="/figures/" />
    <meta-data android:name="books" android:value="@string/books" />
</provider>
```

The string android:authorities lists the providers exposed by an application. For example, if the URI of a provider is content://com.example.android.books. contentprovider/wishlist/English, content:// is the scheme, com.example. android.books.contentprovider is the authority, and wishlist/English is the path. At least one authority has to be specified. Semicolons should separate multiple authorities. It should follow Java namespace rules to avoid conflicts.

The boolean android:enabled tag specifies that the system can initiate the provider. If the value is true, the system can. A value false does not let the system initiate the provider. It is important to note that both the android:enabled attributes, one in the <application> tag and the other in the <provider> tag, need to be true for this to happen.

If the provider is published to other applications, android:exported is set to true. The default value is true for applications with android:targetSdkVersion or android:minSdkVersion set to 16 or lower. For all other applications, the default value is false.

The attribute tag android:grantUriPermissions is used to provide one time access to data that is protected by permissions otherwise and is not accessible by the component. This facility, if set to true, lets the component overcome the restrictions imposed by the android:readPermission, android:writePermission, and android:permission attributes and will allow access to any of Content Provider's data. If this attribute is set to false then permissions can only be granted to datasets listed in the <grant-uri-permission> tag. The default value of this tag is false.

The integer android:initOrder is the order in which a provider is initialized. The higher the number, the earlier it is initialized. This attribute is of particular importance if there are dependencies in the providers of an application.

The string android:label is the user-readable label for the Content Provider.

The boolean `android:multiprocess` attribute, if set to true, lets the system create an instance of the provider in each application's process that interacts with it. This avoids the overhead of inter-process communication. The default value is false which means that the provider is instantiated only in the application process that defined it.

The string `android:permission` tag declares the permissions that a client should have to interact with the provider.

The string `android:readPermission` and string `android:writePermission` define permissions that the client should have to read and write provider data respectively. If defined, these permission supersede the `android:permission` value. It is interesting to note that although the string `android:writePermission` allows only writes on the database, it usually uses a WHERE clause and a smart engineer can work around these to read the database. So write permission should be regarded as read permission as well.

The `android:process` attribute defines the process in which the provider should run. Usually, the provider runs in the same process as the application. However, if it is required to run the process in a separate private process, it can be assigned a name starting with a colon (:). If the name begins with a lowercase character, the provider is instantiated in a global process to enable cross application sharing.

The `android:syncable` attribute allows data to sync to the server by setting the value to `true`. A value of `false` does not let data sync to the server.

A `<provider>` tag can contain three sub tags.

The first is `<grant-uri-permission>` with the following syntax:

```
<grant-uri-permission android:path="string"
                      android:pathPattern="string"
                      android:pathPrefix="string" />
```

The other is the `<path-permission>` tag with the following syntax:

```
<path-permission android:path="string"
                 android:pathPrefix="string"
                 android:pathPattern="string"
                 android:permission="string"
                 android:readPermission="string"
                 android:writePermission="string" />
```

The third is the `<meta-data>` tag that defines the metadata associated with the provider as follows:

```
<meta-data android:name="string"
           android:resource="resource specification"
           android:value="string" />
```

To provide with provider level single read and write, use `android:readPermission` and `android:writePermission` respectively. To provide blanket provider level read/write permissions, use the `android:permission` attribute. To enable temporary permissions, set the `android:grantUriPermissions` attribute. You can also use the `<grant-uri-permission>` child element for the same. To enable path level permission, use the `<path-permission>` child element of `<provider>`.

Other security consideration

A Content Provider extends the `ContentProvider` abstract class. This class has six methods such as `query()`, `insert()`, `update()`, `delete()`, `getType()`, and `onCreate()`, all of which need to be implemented. If the provider does not support some functionality, an exception should be returned. This exception should be able to communicate across process boundaries.

Synchronization can be an issue if multiple threads are reading and writing provider data. This can be taken care of by making all the previously mentioned methods synchronized by using the keyword `synchronize` so only one thread can access the provider. Alternatively, `android:multiprocess=true` can be set so that an instance is created for each client. Latency and performance issues have to be balanced in this case.

In some cases, to maintain data integrity, data may have to be entered in the provider in a certain format. For example, it might be necessary that a tag append each element. To achieve this, a client may decide to not call the `ContentProvider` and `ContentResolver` classes directly. Instead, an Activity can be entrusted to interface with the provider. All clients who need to access the provider data should send an Intent to this Activity and then this Activity performs the intended action.

SQL injection can easily happen with Content Providers if the value fed to the query is not validated. The following is an example of how it can happen:

```
// mUserInput is the user input
String mSelectionClause =  "var = " + mUserInput;
```

A malicious user can enter any text here. It could be `nothing; DROP TABLE *;`, which will delete tables. Developers should use the same discretion that applies for any SQL query. The user data should be parameterized and vetted for possible bad activities.

The user may decide to use regular expressions to check the syntax of the input that the user enters. The following code snippet shows how to validate user input for alphanumeric characters. The snippet uses the `matches` function of the `String` class.

```
if (myInput.length() <= 0) {
  valid = false;
} else if (!myInput.matches("[a-zA-Z0-9 ]+")) {
  valid = false;
} else {
  valid = true;
}
```

When storing data in the database, you might like to encrypt sensitive information such as passwords and credit card information before storing it. Be aware that encrypting some fields may affect your ability to index and sort fields. Additionally, there are some open source tools, such as SQLCipher for Android (`http://sqlcipher.net`) that provides full SQLite database encryption using 256-bit AES.

Broadcast Receiver

Introduced in API level 1, a Broadcast Receiver is a mechanism for an application to receive Intents from the system or other applications. The beauty of a receiver is that even if the application is not running, it still receives Intents that can trigger further events. The user is unaware of a broadcast. As an example, an application that intends to start a background Service as soon as the system is up can register for the `Intent.ACTION_BOOT_COMPLETE` system Intent. An application that wants to customize itself to a new time zone can register for an `ACTION_TIMEZONE_CHANGED` event. An example of a Service sending out a broadcast Intent is shown in the following figure. Receivers that have registered with the Android system for such a broadcast will receive the broadcast Intent.

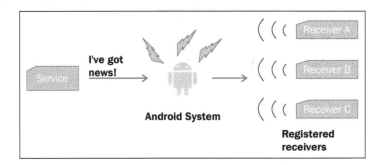

An application can declare a receiver in the manifest file. The receiver class then extends the `BroadcastReceiver` class and implements the `onReceive()` method. Or an application can create and register a receiver dynamically using `Context. registerReceiver`.

Receiver declaration

A receiver can be declared in the manifest file as follows:

```
<receiver android:enabled=["true" | "false"]
          android:exported=["true" | "false"]
          android:icon="drawable resource"
          android:label="string resource"
          android:name="string"
          android:permission="string"
          android:process="string" >
    . . .
</receiver>
```

As an example, let's assume there are two applications. The first application lets users search for books and add books to a wish list. The second application listens for the Intent that a book has been added to wish list. The second application then syncs up the wish list to the list on the server. An example receiver declaration in the manifest file of the second application could be as follows:

```
<receiver
    android:name="com.example.android.book2.MessageListener" >
    <intent-filter>
      <action
        android:name="com.example.android.book1.my-broadcast" />
    </intent-filter>
</receiver>
```

The receiver `com.example.android.book2.MessageListener` is a public receiver and listens to events from application `com.example.android.book1`. The `intent-filter` tag filters out Intents.

The application `book1` can send an Intent as follows:

```
Intent intent = new Intent();
intent.setAction("com.example.android.book1.my-broadcast");
sendBroadcast(intent);
```

The attributes of the `<receiver>` tag are discussed as follows:

- `android:enabled`: Setting this attribute to true lets the system instantiate the receiver. The default value for this attribute is true. This tag has to be used in conjunction with the `android:enabled` attribute of `<application>`. Both have to be true for the system to instantiate it.

- `android:exported`: Setting this attribute to true makes your receiver visible to all applications in the system. If it is false then it can receive Intents only from the same application or applications with the same user ID. If your application does not have Intent Filters then the default value is false as it assumes that this receiver is private to you. If you define Intent filters then the default value is true. In our preceding example, we do have Intent filters, so the receiver is visible to the rest of the system.

- `android:name`: This is the name of the class that implements the receiver. This is a required attribute and should be a fully qualified name of the class. Once you have declared a receiver you should try not to change the name as other applications might be using it and changing the name will break their functionality.

- `android:permission`: You can protect your receiver with permissions. Using this attribute you specify the permissions that the components that send an Intent to your receiver should have. If no permissions are listed here then the permissions of the `<application>` tag are used. If no permissions are specified there as well then your receiver is not protected at all.

- `android:process`: By default the receiver is instantiated in the application process. If you want to, you may declare a name of the process here. If the name starts with a colon (`:`), it is instantiated in a private process within your application. If it starts with a lowercase letter, and your applications have permission to do so, it is run in a global process.

Secure sending and receiving broadcasts

There are two types of broadcasts, normal broadcasts and ordered broadcasts. Normal broadcasts are sent asynchronously using `Context.sendBroadcast()` and all the receivers listening to it will receive it. Ordered broadcasts, sent with a `Context.sendOrderedBoradcast`, are delivered to one receiver at a time. The receiver adds its result and sends it to the next receiver. The order can be set using the `android:priority` attribute in an Intent Filter. If there are multiple filters with same priority, the order in which the broadcast is received is random.

Broadcasts are asynchronous. You send them off but cannot be guaranteed that the receiver will receive it. An application must act gracefully in such a situation.

A broadcast can contain extra information. Any receiver listening to a broadcast can receive a sent broadcast. It is thus prudent to not send any sensitive information in the broadcast. Additionally, broadcasts can be protected with permissions. This is done by supplying a permission string in the sendBroadcast() method. Only applications that have appropriate permissions, by declaring it with <uses-permission> can receive it. Similarly, a permission string can be added to the sendOrderedBroadcast() method.

When a process is still executing onReceive(), it is considered as a foreground process. Once the process is out of the onReceive() method, it is considered as an inactive process and the system will try to kill it. Any asynchronous action being performed in the onReceive() method may be killed. As an example, starting a Service when a broadcast is received should be done using Context.startService().

A sticky broadcast keeps living on until the phone powers off or some component removes it. When the information in the broadcast is updated, the broadcast is updated with the newer information. Any application that has the BROADCAST_STICKY permission can remove or send sticky broadcasts, so do not put any sensitive information in there. Moreover, a sticky broadcast cannot be protected by permissions so they should be used sparingly.

Permissions can be enforced on receivers. As discussed in the previous section, this can be done by adding a permission in the manifest file or dynamically by adding it in the registerReceiver() method.

Starting an **Ice Cream Sandwich**, you can restrict broadcasts to be received by only one application by setting Intent.setPackage.

There are some system broadcast actions that are defined in the Intent class. These events are triggered by the system and an application cannot trigger them. Receivers can register to listen to any of these events. Some of these actions include ACTION_TIMEZONE_CHANGED, ACTION_BOOT_COMPLETED, ACTION_PACKAGE_ADDED, ACTION_PACKAGE_REMOVED, ACTION_POWER_DISCONNECTED, and ACTION_SHUTDOWN.

Local broadcasts

If the broadcast is intended for components within an application only, it is better to use a LocalBroadcastManager helper class. This helper class is a part of the Android support package. Besides being more efficient than sending a global broadcast, it is more secure as it does not leave the application process and other applications cannot see it. A local broadcast does not need to be declared in the manifest, as it is local to the application.

A local broadcast can be created as follows:

```
Intent intent = new Intent("my-local-broadcast");
Intent.putExtra("message", "Hello World!");
LocalBroadcastManager.getInstance(this).sendBroadcast(intent);
```

The following code snippet listens to a local broadcast:

```
@Override
public void onCreate(Bundle savedInstanceState) {
  super.onCreate(savedInstanceState);
  //  ... other code goes here

  LocalBroadcastManager.getInstance(this).registerReceiver(
    mMessageReceiver, new IntentFilter("my-local-broadcast"));
}

private BroadcastReceiver mMessageReceiver = new BroadcastReceiver() {
  @Override
  public void onReceive(Context context, Intent intent) {
    String message = intent.getStringExtra("message");
    Log.d("Received local broadcast" + message);
    // ... other code goes here
  }
};
```

Intents

Intents are Android's mechanism for inter-component communication. Intents are asynchronous so components fire them off and it is the onus of the receiving component to validate the incoming Intent's data and act upon it. Intents are used by the Android system for starting an Activity or Service, for communicating with a Service, to broadcast events or changes, for receiving notifications using pending Intents, and to query the Content Provider.

There are different mechanisms to handle Intents for each component. So, the Intents sent out to Activities, Services, and Broadcast Receivers are only sent to their respective counterparts by the Android system. For example, an event sent out to start an Activity using `Context.startActivity()` will resolve only Activities matching the Intent criterion. Similarly, a broadcast sent out using `Context.sendBroadcast()` will be received only by receivers and not by other components.

Before an Intent is sent out, it is important to check if there is a component to handle the Intent. If there is no component to handle the Intent, the application will crash. Matching Intents can be queried using the `queryIntentActivities()` method of the `PackageManager` class.

 Any rogue application can send an Intent to an exposed component. It is your component's responsibility to validate the input before acting on it.

Intents are basically serialized objects passed between components. This object contains some information used by the other component to act upon. For example, an Activity that logs in the user using their login credentials may start another Activity that loads up the books previously selected by the user using `Context.startActivity()`. In this case, the Intent may contain the user's account name that will be used to fetch books stored on the server.

An `Intent` object contains the following four kinds of information:

1. **Component Name**: A Component Name is required only in case of an explicit Intent. It has to be a fully qualified classname if communicating with an external component or just the classname in case of an internal component.

2. **Action String**: An Action String is the action that should be performed. For example, an Action String `ACTION_ CALL` initiates a phone call. A broadcast action `ACTION_BATTERY_LOW` is a warning to applications about low battery.

3. **Data**: This is the URI of the data along with the MIME type. For example, for `ACTION_CALL`, the data will be of type `tel:`. Both data and the type of data go hand in hand. In order to work on some data, it is important to know the type so that it can be handled appropriately.

4. **Category**: The Category provides additional information about the kind of Intents a component can receive, thereby adding further restrictions. For example, the browser can safely invoke an Activity with a Category of `CATEGORY_BROWSERABLE`.

Intents are asynchronous so no result is expected. In case of Activities, Intents can also be used for starting an Activity for result. This is done using `Context.startActivityForResult()` and the result is returned to the calling Activity using the `finish()` method.

Intents used for broadcasts are usually announcements about an action that just happened. Broadcast Receivers register to listen to such events. Some examples include `ACTION_PACKAGE_ADDED`, `ACTION_TIME_TICK`, `ACTION_BOOT_COMPLETED`. In this scenario, an Intent works like a trigger for some action to be performed once an event takes place.

Do not put any sensitive information in the Intent object. Use another mechanism such as a Content Provider that can be protected by permissions to share information between components.

The receiving component gets extra information attached to the Intent class using getIntent().getExtras(). A secure programming practice requires that this input be validated and vetted for accepted values.

Explicit Intents

A component can send a targeted Intent to only one component. For this to happen, a component should know the fully qualified name of the target component. An Activity in **Application A** sending an explicit Intent to an Activity in **Application B** can be shown graphically as follows:

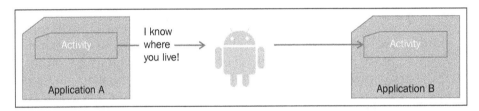

For example, an Activity can explicitly communicate with an internal Activity called ViewBooksActivity using the following code:

```
Intent myIntent = new Intent (this, ViewBooksActivity.class);
startActivity(myIntent);
```

If ViewBooksActivity is an external Activity, the component name should be a fully qualified name of the class. This can be done as follows:

```
Intent myIntent = new Intent (this, "com.example.android.Books.
ViewBooksActivity.class");
startActivity(myIntent);
```

Since intents can be intercepted by any application, if the component name is available, it is best to call the component explicitly.

Implicit Intent

If the fully qualified name of the component is not known, the component can be called implicitly by specifying the action that a receiving component needs to do with it. The system then identifies components that are best suited to handle the Intent by matching the criterion specified in the `Intent` object. An illustration of an implicit Intent is shown as follows. An Activity in **Application A** sends out an Intent, and the system searches the relevant components (based on their Intent Filters and permissions) that can handle such an Intent.

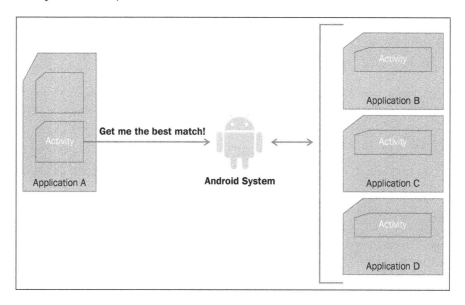

The following are some examples of implicit Intents:

```
// Intent to view a webpage
Intent intent = new Intent(Intent.ACTION_VIEW, Uri.parse("http://www.
google.com"));

// Intent to dial a telephone number
Intent intent = new Intent(Intent.ACTION_DIAL, Uri.
parse("tel:4081112222"));

//Intent to send an email
Intent intent = new Intent(Intent.ACTION_SEND);
emailIntent.setType(HTTP.PLAIN_TEXT_TYPE);
emailIntent.putExtra(Intent.EXTRA_EMAIL, new String[] {"me@example.
com"});
emailIntent.putExtra(Intent.EXTRA_SUBJECT, "Hello Intent!");
emailIntent.putExtra(Intent.EXTRA_TEXT, "My implicit intent");
```

Intent Filter

For a component to be resolved by the system, it needs to be declared in the manifest file with appropriate identifiers. This task is done using the Intent Filters. Intent filters are defined for activities using the `<intent-filter>` subtag of the `<activity>`, `<service>`, or `<receiver>` declaration. While resolving an appropriate Activity for an Intent, the system considers only three aspects of an `Intent` object. These are action, data (both URI and MIME type), and category. All these Intent aspects must match for a successful resolution. A component name is used only for explicit Intents.

Intent Filters must contain the `<action>` subtag and may contain `<category>` and `<data>`. Some examples of the `<intent-filter>` declarations are as follows.

An Activity that is the starting point of the application is identified with these tags:

```
<intent-filter>
  <action android:name="android.intent.action.MAIN" />
  <category android:name="android.intent.category.LAUNCHER" />
</intent-filter>
```

An Activity that lets user request data of the type `book` can be defined as follows:

```
<intent-filter>
  <action android:name="android.intent.action.GET_CONTENT" />
  <category android:name="android.intent.category.DEFAULT" />
  <data android:mimeType="vnd.android.cursor.item/vnd.example.book"/>
</intent-filter>
```

Intent Filters are not security boundaries and should not be relied upon for security. Intent Filters cannot be secured with permissions. Moreover, any component with Intent Filters becomes an exported component and any application can send Intents to this component.

Pending Intent

In case of Intents, the receiving application executes code with its own permission as if it is a part of the receiving application. In case of a pending Intent, the receiving application uses the original application's identity and permissions and executes the code on its behalf.

Thus a pending Intent is a token an application gives to another application so that the other application can execute a piece of code with the original application's permissions and identity. A pending Intent will execute even if the sending application process is killed or destroyed. This property of pending Intents can be used beautifully to send notification to the originating application once an event has happened. Pending Intents can be either explicit or implicit.

For additional security, so that only one component receives the Intent, a component can be baked into the Intent by using the setComponent() method. By default, a pending Intent cannot be modified by the receiving component. This is good for security reasons. The only part that the receiving component can edit is extras. The sender can, however, set flags to explicitly enable receiving components to edit PendingIntent. For this to happen, the sender sets rules for using the fillIn(Intent, int) method. For example, if the sender wants to let the receiver overwrite the data field, even if it is set, then the sender can set FILL_IN_DATA=true. This is a very sensitive operation and should be done with care.

Summary

In this chapter, we reviewed the four components of an Android system—Activities, Services, Content Providers, and Broadcast Receivers, and inter-component communication mechanisms—Intents and Binders. Security begins with secure declaration of these components. As is the general rule with security, exposing the minimum is always a good idea. All Android components are protected by permissions. Intents are asynchronous components and should always validate their input. Intent Filters are a good way to reduce the attack surface of an application, but an explicit Intent can still send Intents to it. Now that we understand the Android components and communication mechanism, let's move on to the next chapter to review Android permissions in detail.

3
Permissions

Permissions are the focus of this chapter. They are an integral part of an Android application and almost all application developers and users will encounter them at one time or the other. As we discussed in *Chapter 1, Android Security Model – the Big Picture*, install time application review is the most important security gate. This step is an all or nothing decision by the user; a user either accepts all the listed permissions or declines to download the app. So, as a user of an Android phone, it is important to understand permissions to make prudent decisions about which application to install. Permissions form the basis for securing components and protecting user data.

This chapter begins with an introduction of an existing permission in the Android system. We discuss the four permission protection levels namely Normal, Dangerous, Signature, and SignatureOrSystem. Then, we will discuss about how to secure an application and its components using permissions. Next, we learn how to extend the permission model by adding user defined permissions. This section will discuss permission groups, permission trees, and the syntax to create a new permission in the manifest file.

Permission protection levels

At the application level, Android security is based on permissions. Using this permission-based model, the Android system protects system resources, such as camera and Bluetooth, and application resources, such as files and components. An application should have privileges to act upon or use these resources. Any application intending to use these resources needs to declare to the user that it will be accessing these resources. For example, if an application will be sending and reading a SMS, it will need to declare `android.permission.SEND_SMS` and `android.permission.READ_SMS` in the manifest file.

Permissions are also an effective method for access control between applications. An application's manifest file contains a list of permissions. Any external application that wishes to access this application's resources should possess these permissions. This is discussed in greater detail in the next chapter.

All Android permissions are declared as constants in the `Manifest.permission` class. However, this class does not mention the type of permission. This can be used to check the Android source code. I have tried to list some of these permissions in the following sections. The list of permissions keeps changing based on the functionality, so it is best to refer to the Android source code for an up-to-date listing of permissions. For example, `android.permission.BLUETOOTH` has been around since API level 1 but `android.permission.AUTHENTICATE_ACCOUNTS` was added in API 5. You can get information to get the Android source code at `source.android.com`.

All Android permissions lie in one of the four protection levels. Permissions of any protection levels need to be declared in the manifest file. Third party apps can only use permissions with protection level 0 and 1. These protection levels are discussed as follows:

- Normal permissions: Permissions in this level (level 0) cannot do much harm to the user. They generally do not cost users money, but they might cause users some annoyance. When downloading an app, these permissions can be viewed by clicking on the **See All** arrow. These permissions are automatically granted to the app. For example, permissions `android.permission.GET_PACKAGE_SIZE` and `android.permission.FLASHLIGHT` lets the application get the size of any package and access a flashlight respectively.

  The following is a list of some of the normal permissions that exist in the Android system at the time of writing the book.

  Permissions that are used to set user preferences include:

  - `android.permission.EXPAND_STATUS_BAR`
  - `android.permission.KILL_BACKGROUND_PROCESSES`
  - `android.permission.SET_WALLPAPER`
  - `android.permission.SET_WALLPAPER_HINTS`
  - `android.permission.VIBRATE`
  - `android.permission.DISABLE_KEYGUARD`
  - `android.permission.FLASHLIGHT`

Permissions that allow user to access system or application information include:

- ° `android.permission.ACCESS_LOCATION_EXTRA_COMMANDS`
- ° `android.permission.ACCESS_NETWORK_STATE`
- ° `android.permission.ACCESS_WIFI_STATE`
- ° `android.permission.BATTERY_STATS`
- ° `android.permission.GET_ACCOUNTS`
- ° `android.permission.GET_PACKAGE_SIZE`
- ° `android.permission.READ_SYNC_SETTINGS`
- ° `android.permission.READ_SYNC_STATS`
- ° `android.permission.RECEIVE_BOOT_COMPLETED`
- ° `android.permission.SUBSCRIBED_FEEDS_READ`
- ° `android.permission.WRITE_USER_DICTIONARY`

Permissions that users should be asked for carefully include `android.permission.BROADCAST_STICKY`, that lets an application send a sticky broadcast which stays alive even after the broadcast has been delivered.

- Dangerous permissions: Permissions in this protection level (level 1) are always shown to the user. Granting dangerous permissions to apps allow them to access device features and data. These permissions cause user privacy or financial loss. For example, granting dangerous permissions, such as `android.permission.ACCESS_FINE_LOCATION` and `android.permission.ACCESS_COARSE_LOCATION`, lets an app access the user's location that might be a privacy concern if the app does not need such a capability. Similarly, granting an app `android.permission.READ_SMS` and `android.permission.SEND_SMS` permission lets an application send and receive SMS which might be a privacy issue.

At any given point, a user can check the permissions granted to any application by going to the settings and selecting the application. Refer to the following two images that show the permissions for the Gmail application. The first image shows the dangerous permissions that are always displayed to the user. Notice the drop-down menu button **Show All**. This option shows all the permissions requested by the application. Notice the permission, **Hardware controls**, and a normal permission that is not displayed to the user by default.

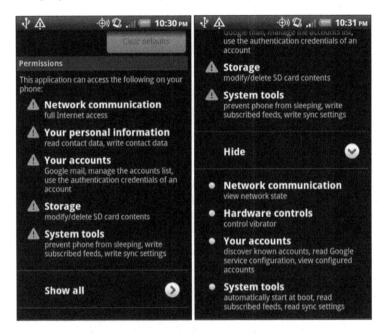

The following is a list of some of the dangerous permissions in the Android system at the time of writing the book.

Some dangerous permissions can be costly for the users. For example, an application that sends a SMS or subscribes to paid feeds can cause users huge bucks. The following are some other examples:

- ° `android.permission.RECEIVE_MMS`
- ° `android.permission.RECEIVE_SMS`
- ° `android.permission.SEND_SMS`
- ° `android.permission.SUBSCRIBED_FEEDS_WRITE`

Permissions that have the power to change the state of the phone include the following. These should be used carefully, as they can make the system unstable, cause annoyance, and can also make the system less secure. For example, permission to mount and unmount filesystems can change the state of the phone. Any malicious application with the permission to record audio can easily eat up phone's memory with garbage audio. The following are some examples:

- `android.permission.MODIFY_AUDIO_SETTINGS`
- `android.permission.MODIFY_PHONE_STATE`
- `android.permission.MOUNT_FORMAT_FILESYSTEMS`
- `android.permission.WAKE_LOCK`
- `android.permission.WRITE_APN_SETTINGS`
- `android.permission.WRITE_CALENDAR`
- `android.permission.WRITE_CONTACTS`
- `android.permission.WRITE_EXTERNAL_STORAGE`
- `android.permission.WRITE_OWNER_DATA`
- `android.permission.WRITE_SETTINGS`
- `android.permission.WRITE_SMS`
- `android.permission.SET_ALWAYS_FINISH`
- `android.permission.SET_ANIMATION_SCALE`
- `android.permission.SET_DEBUG_APP`
- `android.permission.SET_PROCESS_LIMIT`
- `android.permission.SET_TIME_ZONE`
- `android.permission.SIGNAL_PERSISTENT_PROCESSES`
- `android.permission.SYSTEM_ALERT_WINDOW`

Some dangerous permissions can be privacy risks. Permissions that let users read SMS, logs, and calendar can be easily used by botnets and Trojans to do interesting stuff on a command of their remote owners. The following are some other examples:

- `android.permission.MANAGE_ACCOUNTS`
- `android.permission.MODIFY_AUDIO_SETTINGS`
- `android.permission.MODIFY_PHONE_STATE`
- `android.permission.MOUNT_FORMAT_FILESYSTEMS`

- ○ android.permission.MOUNT_UNMOUNT_FILESYSTEMS
- ○ android.permission.PERSISTENT_ACTIVITY
- ○ android.permission.PROCESS_OUTGOING_CALLS
- ○ android.permission.READ_CALENDAR
- ○ android.permission.READ_CONTACTS
- ○ android.permission.READ_LOGS
- ○ android.permission.READ_OWNER_DATA
- ○ android.permission.READ_PHONE_STATE
- ○ android.permission.READ_SMS
- ○ android.permission.READ_USER_DICTIONARY
- ○ android.permission.USE_CREDENTIALS

- **Signature permissions**: Permissions in this protection level (level 2) allow two applications authored by the same developer, access each other's components. This permission is automatically granted to the app if the app being downloaded has the same certificate as the application that declared the permission. For example, application A defines a permission com.example.permission.ACCESS_BOOK_STORE. Application B, signed by the same certificate as application A, declares it in its manifest file. Since both application A and B have the same certificate, this permission is not shown to the user when installing the application B. A user can certainly view it, using **See All**. An app can perform really powerful actions with this permission of this group. For example, with android.permission. INJECT_EVENTS, an app can inject events such as keys, touch, and trackball into any application and android.permission.BROADCAST_SMS can broadcast an SMS acknowledgement. This permission defined by the Android systems that lie in this protection group is reserved for system apps only.

Some permissions in this level allow applications to use system level features. For example, ACCOUNT_MANAGER permission lets the applications use account authenticators and BRIK permissions allow the applications to brick the phone. The following is a list of some of the signature permissions at the time of writing the book. For a complete reference check the Android source code or the Manifest.permission class:

- ○ android.permission.ACCESS_SURFACE_FLINGER
- ○ android.permission.ACCOUNT_MANAGER
- ○ android.permission.BRICK
- ○ android.permission.BIND_INPUT_METHOD

- ○ android.permission.SHUTDOWN
- ○ android.permission.SET_ACTIVITY_WATCHER
- ○ android.permission.SET_ORIENTATION
- ○ android.permission.HARDWARE_TEST
- ○ android.permission.UPDATE_DEVICE_STATS
- ○ android.permission.CLEAR_APP_USER_DATA
- ○ android.permission.COPY_PROTECTED_DATA
- ○ android.permission.CHANGE_COMPONENT_ENABLED_STATE
- ○ android.permission.FORCE_BACK
- ○ android.permission.INJECT_EVENTS
- ○ android.permission.INTERNAL_SYSTEM_WINDOW
- ○ android.permission.MANAGE_APP_TOKENS

Some permissions in this level allow applications to send system level broadcasts and intents such as broadcasting intents and SMS. These permissions include:

- ○ android.permission.BROADCAST_PACKAGE_REMOVED
- ○ android.permission.BROADCAST_SMS
- ○ android.permission.BROADCAST_WAP_PUSH

Other permissions in this level allow applications to access system level data that third party applications do not have. These permissions include:

- ○ android.permission.PACKAGE_USAGE_STATS
- ○ android.permission.CHANGE_BACKGROUND_DATA_SETTING
- ○ android.permission.BIND_DEVICE_ADMIN
- ○ android.permission.READ_FRAME_BUFFER
- ○ android.permission.DEVICE_POWER
- ○ android.permission.DIAGNOSTIC
- ○ android.permission.FACTORY_TEST
- ○ android.permission.FORCE_STOP_PACKAGES
- ○ android.permission.GLOBAL_SEARCH_CONTROL

- SignatureOrSystem permissions: As with signature protection level, this permission is granted to applications with the same certificate as the application that defined the permission. In addition, this protection level includes applications with the same certificate as the Android system image. This permission level is mainly used for applications that are built by handset manufacturers, carriers, and system applications. These permissions are not allowed for third party apps. These permissions let apps perform some very powerful functions. For example, the permission `android.permission.REBOOT` allows an app to reboot the device. The permission `android.permission.SET_TIME` lets an app set system time.

A list of some of the SignatureOrSystem permissions as of the time of writing the book is as follows:

  - `android.permission.ACCESS_CHECKIN_PROPERTIES`
  - `android.permission.BACKUP`
  - `android.permission.BIND_APPWIDGET`
  - `android.permission.BIND_WALLPAPER`
  - `android.permission.CALL_PRIVILEGED`
  - `android.permission.CONTROL_LOCATION_UPDATES`
  - `android.permission.DELETE_CACHE_FILES`
  - `android.permission.DELETE_PACKAGES`
  - `android.permission.GLOBAL_SEARCH`
  - `android.permission.INSTALL_LOCATION_PROVIDER`
  - `android.permission.INSTALL_PACKAGES`
  - `android.permission.MASTER_CLEAR`
  - `android.permission.REBOOT`
  - `android.permission.SET_TIME`
  - `android.permission.STATUS_BAR`
  - `android.permission.WRITE_GSERVICES`
  - `android.permission.WRITE_SECURE_SETTINGS`

Application level permissions

There are two ways to apply permissions to the entire application. In the first case, an application declares what permissions the application requires to function properly. So, an application that will be sending out SMS will declare such permission in the manifest file. In the second case, the application can declare what permissions other applications trying to interact with this application should have. For example, an application can declare that any application that wants to interact with one of its components should have permissions to access the camera. Both these kinds of permissions have to be declared in the manifest file. Let us go through them one by one.

This `<uses-permission>` tag is declared inside `<manifest>` and declares what permissions the application requests to function properly. The syntax of the tag is the following:

```
<uses-permission android:name=" " />
```

The user, when downloading the application, has to accept these permissions. `android:name` is the name of the permission. An example declaration of this tag is as follows. The following permission declares that the application that the user is about to install would access the user's SMS:

```
<uses-permission android:name="android.permission.READ_SMS"/>
```

The `<application>` tag has an attribute called `android:permission` that declares blanket permissions for components. These are the permissions any application trying to interact with this application's components need to have. This is shown in the following code. The following code shows that applications interacting with any component of `MyApplication` should have permission to access the camera:

```
<application android:name="MyApplication" android:icon="@drawable/
icon" android:label="@string/app_name""android.permission.CAMERA"/>
```

As discussed in the next section, individual components can set permissions as well. Component permissions override the permission set using the `<application>` tag. The preceding method is the best place to declare the blanket permissions for all components.

Component level permissions

All Android components can be secured using permissions. The following figure illustrates this concept:

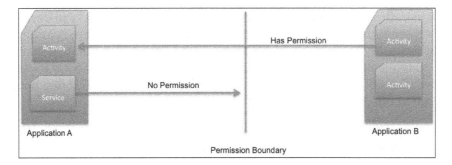

Let's talk about permission declaration and enforcement for each component.

Activity

Any Activity can be secured by permission, by calling out the permission in Activity declaration in the `<activity>` tag. For example, the Activity `OrderActivity` with a custom permission `com.example.project.ORDER_BOOK` will be declared as follows. Any component that tries to launch `OrderActivity` needs to have this custom permission.

```
<activity android:name=".OrderActivity" android:permission="com.
example.project.ORDER_BOOK" android:exported="false"/>
```

In case of activities, permission enforcement happens when launching the Activity, by using `Context.startActivity()` and `Context.startActivityForResult()`. In case the launching component does not have appropriate permissions, a `SecurityException` is thrown.

Service

Any Service can be protected using permission by listing out the required permissions in the `<service>` tag. For example, the Service `FindUsefulSMS` that identifies the SMS based on keywords declares a permission `android.permission.READ_SMS`. This permission will be declared as follows. Any component that tries to launch `FindUsefulSMS` needs to have this permission.

```
<service android:name=".FindUsefulSMS" android:enabled="true"
android:permission="android.permission.READ_SMS">
</service>
```

Permission enforcement for a Service is done at the time of launching a Service by using `Context.startService()`, stopping a Service by using `Context.stopService()`, and binding to a Service by using `Context.bindService()`. In case the requesting component does not have the appropriate permissions, a `SecurityException` is thrown.

If the Service exposes a binder interface that other applications can bind to, caller permissions can be checked when binding to the binder by using `Context.checkCallingPermission()`.

Content Provider

Content providers can be secured with permissions specified in the `<provider>` tag. In the following example, any component that wants to talk to the provider should have the `android.permission.READ_SMS` permission:

```
<provider
    android:authorities="com.example.android.books.contentprovider"
    android:name=".contentprovider.MyBooksdoContentProvider"
    android:grantUriPermissions="true"
    android:Permission="android.permission.READ_CALENDAR"/>
```

As discussed in *Chapter 2, Application Building Blocks*, the `<provider>` tag also has fine-grained read and write permission attributes. To be able to read from a `<provider>` tag, the application should have the read permission. This is checked during `ContentResolver.query()`. To be able to update, delete, and insert into a provider, a component should have read and write permissions. These permissions are checked during `ContentResolver.insert()`, `ContentResolver.update()`, and `ContentResolver.delete()`. Failure to have appropriate permission will result in `SecurityException` being thrown by the call.

The `<grant-uri-permission>` tag is a child of the `<provider>` tag and is used to grant access to some specific data sets of a provider for a limited amount of time. Consider an example of an app that saves SMS to a database. Some SMSs may have a photo attached to it. For an application to view the SMS correctly, it will launch the image viewer, which might not have access to the provider. The **URI (Universal Resource Identifier)** permissions will let the image viewer have read permission for that particular picture. In the previous example, where the provider set `android:grantIriPermissions="true"`, the image viewer will have read permission to the entire provider. This poses a security risk. To give limited access, the provider can declare which parts of the provider it wants to be open to the URI permissions.

The syntax of the URI permission is as follows:

```
<grant-uri-permission android:path="string"
                       android:pathPattern="string"
                       android:pathPrefix="string" />
```

 The URI permissions are not recursive.

What I find most interesting is that we can use wildcards and patterns to define which parts of the provider we want to be able to enforce the URI permissions to. An example of this is as follows:

```
<grant-uri-permission android:pathPattern="/picture/" />
```

Remember to revoke the URI permissions once the task is completed by using `Context.revokeUriPermission()`.

Broadcast receiver

Broadcasts can be secured with permissions in two ways. In the first case, the receiver protects itself with permissions so it receives only the broadcasts it wants to hear. In the other case, a broadcaster selects which receivers can receive the broadcast. We will discuss both the scenarios in the following section.

Any receiver can be secured by calling out the permission in the receiver declaration in the `<receiver>` tag. For example, the receiver `MyListener` declares a permission `android.permission.READ_SMS` and this will be declared as follows. `MyListener` will receive broadcasts only from broadcasters with the permission `android.permission.READ_SMS`.

```
<receiver android:name=".MyListener"
android:permission="android.permission.READ_SMS">
        <intent-filter>
            <action android:name=
                "android.provider.Telephony.SMS_RECEIVED" />
        </intent-filter>
</receiver>
```

 Remember, sticky broadcasts cannot be protected by permissions.

The required permissions for receiving a broadcast are checked after the broadcast intent is delivered, that is, after the call to `Context.sendBroadcast()` returns. So, no exception is thrown if the broadcaster does not have the appropriate permissions; just that the broadcast will not be delivered. If the receiver is dynamically created by using `Context.registerReceiver()`, the permission can be set when creating this receiver.

The second case, where a broadcaster restricts which receivers can receive an intent, is done by using the method `sendBroadcast()`. An example of a broadcast that will only be sent to receivers of applications with the `android.permission.READ_SMS` permission is defined in the following code snippet:

```
Intent intent = new Intent();
intent.setAction(MY_BROADCAST_ACTION);
sendBroadcast(intent, "android.provider.Telephony.SMS_RECEIVED");
```

 Permissions declared with components are not granted to the application. They are permissions that an application whose component is trying to interact with it should have.

Extending Android permissions

Developers can extend the permission system by adding their own permissions. These permissions will be displayed to the user during the time of downloading the app, so it is important that they are localized and labeled properly.

Adding a new permission

Developers may choose to add just a new permission or an entire tree of permissions. Declaring new permissions is done in the manifest file. To add a new permission, an application can declare it by using the `<permission>` tag as shown in the following code snippet:

```
<permission android:name="string"
            android:description="string resource"
            android:icon="drawable resource"
            android:label="string resource"
            android:permissionGroup="string"
            android:protectionLevel=["normal" | "dangerous" |
                       "signature" | "signatureOrSystem"] />
```

A description of the attributes used in the preceding code snippet for the new permission group is as follows:

- `android:name`: This is the name of the new permission being declared.

- `android:description`: This describes the new permission being declared in detail.

- `android:icon`: This is the permission icon.

- `android:label`: This is the label displayed to the user at install time.

- `android:permissionGroup`: This assigns a permission to a pre-existing user defined group or a new group. If no name is specified, permission does not belong to any group, which is also fine. I will discuss how to create a permission group later in this section.

- `android:protectionLevel`: This specifies the protection level of the new permission. These protection levels have been discussed earlier in this chapter.

An example for such permission can be as follows:

```
<permission android:name="com.example.android.book.READ_BOOKSTORE"
        android:description="@string/perm_read_bookstore"
        android:label="Read access to books database"
        android:permissionGroup="BOOKSTORE_PERMS"
        android:protectionLevel="dangerous"/>
```

To account for localization and maintenance, it is always better to use a string resource than a raw string.

Once you declare a new permission, make sure to declare it in the `<uses-permission>` tag.

Creating a permission group

A permission group can be created by using a `<permission-group>` tag. It is a logical grouping of permissions and when presenting them to the user, they are all presented together. A permission group is created by using the following syntax:

```
<permission-group android:name="string"
        android:description="string resource"
                android:icon="drawable resource"
                android:label="string resource" />
```

A description of the attributes used in the preceding code snippet for the new permission group is as follows:

- android:name: This is the name of the new permission group. This is the name mentioned in the <permission> tag.

- android:description: This describes the new permission group being declared in detail.

- android:icon: This is the permission group icon.

- android:label: This is the label to be displayed at the time of install.

An example of a permission group with permissions for the bookstore can be declared as follows:

```
<permission-group android:description="@string/perm_group_bookstore"
                  android:label="@string/perm_group_bookstore_label"
                  android:name="BOOKSTORE_PERMS" />
```

Creating a permission tree

If there is a need to arrange permissions as a namespace, such that a permission tree can be created, then the application can declare a <permission-tree> tag. An example of such a tree is as follows:

```
com.example.android.book
com.example.android.book.READ_BOOK
com.example.android.book.bookstore.READ_BOOKSTORE
com.example.android.book.bookstore.WRITE_BOOKSTORE
```

This tag does not define any new permissions, it just creates a namespace for you to group permissions. I see this concept being used by developers that have multiple applications and all such applications talk to each other. The syntax for a <permission-tree> tag is defined as follows:

```
<permission-tree android:name="string"
    android:icon="drawable resource"
                android:label="string resource"  />
```

A description of the attributes used in the preceding code snippet for the new permission group is as follows:

- `android:name`: This is the name of the new permission group. The name should have at least three segments separated by a period, for example, `com.example.android` is fine, but `com.example` is not.

- `android:icon`: This is the permission group icon.

- `android:label`: This is the label to be displayed to the user at the time of install.

An example declaration is as follows:

```
<permission-tree android:name="com.example.android.book"
                 android:label="@string/perm_tree_book"  />
```

Summary

Permissions are the core of Android application security and this chapter covered permissions in detail. We learned the four permission protection levels, how to protect the components with permissions, and how to define new permissions. Awareness and understanding of the permission model is essential both for developers and users of an Android phone. Now that we are armed with the knowledge of components, inter-component communication and permissions, let's march to the next chapter and learn how to define an application's policy file.

4
Defining the Application's Policy File

This chapter brings together all the learning we have done so far. We will use the application components, Intents, and permissions and put them all together to define our application's policy file. This policy file is called `AndroidManifest.xml` and is by far the most important file of an application. As you will see, this file is the place to define access control policy for your application and components. This is also the place to define application and component level specifics that the Android system will use to interact with your application.

The chapter begins with a discussion of `AndroidManfiest.xml`. We will discuss the two important tags: `<manifest>` and `<application>` that we have not discussed so far. Next, we will discuss the actions that we can perform in the manifest file such as declaring permission, sharing a process with other applications, external storage, and managing component visibility. The chapter closes with a discussion of a checklist for your policy file before you release your application. You should adapt the checklist according to your use case. Once you have a comprehensive checklist, you can refer to it every time you are ready to make a new release.

The AndroidManifest.xml file

All Android applications need to have a manifest file. This file has to be named as `AndroidManifest.xml` and has to be placed in the application's root directory. This manifest file is the application's policy file. It declares the application components, their visibility, access rules, libraries, features, and the minimum Android version that the application runs against.

The Android system uses the manifest file for component resolution. Thus, the `AndroidManfiest.xml` file is by far the most important file in the entire application, and special care is required when defining it to tighten up the application's security.

The manifest file is not extensible, so applications cannot add their own attributes or tags. The complete list of tags with how these tags can be nested is as follows:

```
<uses-sdk><?xml version="1.0" encoding="utf-8"?>

<manifest>
    <uses-permission />
    <permission />
    <permission-tree />
    <permission-group />
    <instrumentation />
    <uses-sdk />
    <uses-configuration />
    <uses-feature />
    <supports-screens />
    <compatible-screens />
    <supports-gl-texture />
    <application>
      <activity>
          <intent-filter>
              <action />
              <category />
              <data />
          </intent-filter>
          <meta-data />
      </activity>
      <activity-alias>
          <intent-filter>        </intent-filter>
          <meta-data />
      </activity-alias>
      <service>
          <intent-filter>        </intent-filter>
          <meta-data/>
      </service>
      <receiver>
          <intent-filter>        </intent-filter>
          <meta-data />
      </receiver>
      <provider>
          <grant-uri-permission />
          <meta-data />
          <path-permission />
      </provider>
      <uses-library />
    </application>
</manifest>
```

We have covered most of the tags in previous chapters.

Only two tags, `<manifest>` and `<application>`, are the required tags. There is no specific order to declare components.

The `<manifest>` tag declares the application specific attributes. It is declared as follows:

```
<manifest xmlns:android="http://schemas.android.
  com/apk/res/android"
          package="string"
          android:sharedUserId="string"
          android:sharedUserLabel="string resource"
          android:versionCode="integer"
          android:versionName="string"
          android:installLocation=["auto" | "internalOnly" |
             "preferExternal"] >

</manifest>
```

An example of the `<manifest>` tag is shown in the following code snippet. In this example, the package is named `com.android.example`, the internal version is 10, and the user sees this version as 2.7.0. The install location is decided by the Android system based on where it has room to store the application.

```
<manifest package="com.android.example"
   android:versionCode="10"
   android:versionName="2.7.0"
   android:installLocation="auto"
   xmlns:android="http://schemas.android.com/apk/res/android">
```

The attributes of the `<manifest>` tag are as follows:.

- `package`: This is the name of the package. This is the Java style namespace of your application, for example, `com.android.example`. This is the unique ID of your application. If you change the name of a published application, it is considered a new application and auto updates will not work.

- `android:sharedUserId`: This attribute is used if two or more applications share the same Linux ID. This attribute is discussed in detail in a later section.

- `android:sharedUserLabel`: This is the user readable name of the shared user ID and only makes sense if `android:sharedUserId` is set. It has to be a string resource.

- `android:versionCode`: This is the version code used internally by the application to track revisions. This code is referred to when updating an application with the more recent version.

- android:versionName: This is the version of the application shown to the user. It can be set as a raw string or as a reference, and is only used for display to users.

- android:installLocation: This attribute defines the location where an APK will be installed. This attribute is discussed in detail later in the chapter.

The application tag is defined as follows:

```
<application android:allowTaskReparenting=["true" | "false"]
             android:backupAgent="string"
             android:debuggable=["true" | "false"]
             android:description="string resource"
             android:enabled=["true" | "false"]
             android:hasCode=["true" | "false"]
             android:hardwareAccelerated=["true" | "false"]
             android:icon="drawable resource"
             android:killAfterRestore=["true" | "false"]
             android:largeHeap=["true" | "false"]
             android:label="string resource"
             android:logo="drawable resource"
             android:manageSpaceActivity="string"
             android:name="string"
             android:permission="string"
             android:persistent=["true" | "false"]
             android:process="string"
             android:restoreAnyVersion=["true" | "false"]
             android:supportsRtl=["true" | "false"]
             android:taskAffinity="string"
             android:theme="resource or theme"
             android:uiOptions=["none" |
                 "splitActionBarWhenNarrow"] >

</application>
```

An example of the `<application>` tag is shown in the following code snippet. In this example, the application name, description, icon, and label are set. The application is not debuggable and the Android system can instantiate the components.

```
<application android:label="@string/app_name"
    android:description="@string/app_desc"
    android:icon="@drawable/example_icon"
    android:enabled="true"
    android:debuggable="false">

</application>
```

Many attributes of the <application> tag serve as the default values for the components declared within the application. These tags include permission, process, icon, and label. Other attributes such as debuggable and enabled are set for the entire application. The attributes of the <application> tag are discussed as follows:

- android:allowTaskReparenting: This value can be overridden by the <activity> element. It allows an Activity to re-parent with the Activity it has affinity with, when it is brought to the foreground.

- android:backupAgent: This attribute contains the name of the backup agent for the application.

- android:debuggable: This attribute when set to true allows an application to be debugged. This value should always be set to false before releasing the app in the market.

- android:description: This is the user readable description of an application set as a reference to a string resource.

- android:enabled: This attribute if set to true, the Android system can instantiate application components. This attribute can be overridden by components.

- android:hasCode: This attribute if set to true, the application will try to load some code when launching thecomponents.

- android:hardwareAccelerated: This attribute when set to true allows an application to support hardware accelerated rendering. It was introduced in the API level 11.

- android:icon: This is the application icon as a reference to a drawable resource.

- android:killAfterRestore: This attribute if set to true, the application will be terminated once its settings are restored during a full-system restore.

- android:largeHeap: This attribute lets the Android system create a large Dalvik heap for this application and increases the memory footprint of the application, so this should be used sparingly.

- android:label: This is the user readable label for the application.

- android:logo: This is the logo for the application.

- android:manageSpaceActivity: This value is the name of the Activity that manages the memory for the application.

- android:name: This attribute contains the fully qualified name of the subclass that will be instantiated before any other component is started.

- android:permission: This attribute can be overridden by a component and sets the permission that a client should have to interact with the application.

- `android:persistent`: Usually used by a system application, this attribute allows the application to be running all the time.

- `android:process`: This is the name of the process in which a component will run. This can be overridden by any component's `android:process` attribute.

- `android:restoreAnyVersion`: This attribute lets the backup agent attempt a restore even if the backup currently stored is by a newer application than what is attempting to restore now.

- `android:supportsRtl`: This attribute when set to `true` supports right-to-left layouts. It was added in the API level 17.

- `android:taskAffinity`: This attribute lets all activities have affinity with the package name, unless it is set by the Activity explicitly.

- `android:theme`: This is a reference to the style resource for the application.

- `android:uiOptions`: This attribute if set to `none`, there are no extra UI options; if set to `splitActionBarWhenNarrow`, a bar is set at the bottom if constrained for the screen.

In the following sections we will discuss how to handle specific requirements using the policy file.

Application policy use cases

This section discusses how to define the application policies using the manifest file. I have used use cases and we will discuss how to implement these use cases in the policy file.

Declaring application permissions

An application on the Android platform has to declare what resources it intends to use for proper functioning of the application. These are the permissions that are displayed to the user when they download the application. As discussed in *Chapter 3, Permissions*, an application can define custom permissions as well. Application permissions should be descriptive so that users can understand them. Also, as is the general rule with security, it is important to request the minimum permissions required.

Application permissions are declared in the manifest file by using the tag `<uses-permission>`. An example of a location-based manifest file that uses the GPS for retrieving location is shown in the following code snippet:

```
<uses-permissionandroid:name="android.
    permission.ACCESS_COARSE_LOCATION" />
<uses-permissionandroid:name="android.
```

```
    permission.ACCESS_FINE_LOCATION" />
<uses-permissionandroid:name="android.
    permission.ACCESS_LOCATION_EXTRA_COMMANDS" />
<uses-permissionandroid:name="android.
    permission.ACCESS_MOCK_LOCATION" />
<uses-permissionandroid:name="android.permission.INTERNET" />
```

These permissions will be displayed to the users when they install the app, and can always be checked by going to **Application** under the settings menu. These permissions are seen in the following screenshot:

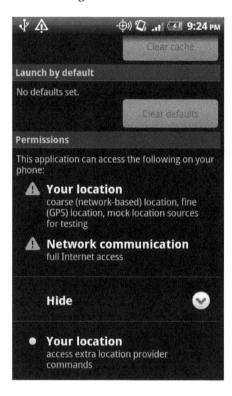

Declaring permissions for external applications

The manifest file also declares the permissions an external application (which does not run with the same Linux ID) needs to access the application components. This can be one of two places in the policy file: in the `<application>` tag or along with the component in the `<activity>`, `<provider>`, `<receiver>`, and `<service>` tag.

If there are permissions that all components of an application require, then it is easy to specify them in the `<application>` tag. If a component requires some specific permission, then those can be defined in the specific component tag. Remember, only one permission can be declared in any of the tags. If a component is protected by permission then the component permission overrides the permission declared in the `<application>` tag.

The following is an example of an application that requires external applications to have `android.permission.ACCESS_COARSE_LOCATION` to access its components and resources:

```
<application
    android:allowBackup="true"
    android:icon="@drawable/ic_launcher"
    android:label="@string/app_name"
    android:permission="android.
        permission.ACCESS_COARSE_LOCATION">
```

If a Service requires that any application component that accesses it should have access to the external storage, then it can be defined as follows:

```
<service android:enabled="true"
    android:name=".MyService"
        android:permission="android.
            permission.WRITE_EXTERNAL_STORAGE">
</service>
```

If a policy file has both the preceding tags then when an external component makes a request to this Service, it should have `android.permission.WRITE_EXTERNAL_STORAGE`, as this permission will override the permission declared by the application tag.

Applications running with the same Linux ID

Sharing data between applications is always tricky. It is not easy to maintain data confidentiality and integrity. Proper access control mechanisms have to be put in place based on who has access to how much data. In this section, we will discuss how to share application data with the internal applications (signed by the same developer key).

Android is a layered architecture with an application isolation enforced by the operating system itself. Whenever an application is installed on the Android device, the Android system gives it a unique user ID defined by the system. Notice that the two applications, **example1** and **example2**, in the following screenshot are the applications run as separate user IDs, **app_49** and **app_50**:

```
⊙ ○ ○                    platform-tools — adb — 80×9
app_33     491   37     159244 31092 ffffffff 400113c0 S com.android.quicksearchbo
x
app_49     521   37     157524 32428 ffffffff 400113c0 S com.example.example1
root       535   46     704    324   c003d800 4000d284 S /system/bin/sh
root       537   535    660    348   c0099f1c 400107b4 S logcat
root       538   46     704    332   c003d800 4000d284 S /system/bin/sh
app_50     572   37     159020 33872 ffffffff 400113c0 S com.example.example2
root       586   538    900    348   00000000 40010458 R ps
#
```

However, an application can request the system for a user ID of its choice. The other application can then request the same user ID as well. This creates tight coupling and does not require components to be made visible to the other application or to create shared content providers. This kind of tight coupling is done in the manifest tags of all applications that want to run in the same process.

The following is a snippet of manifest files of the two applications com.example. example1 and com.example.example2 that use the same user ID:

```
<manifest xmlns:android="http://schemas.android.
  com/apk/res/android"
    package="com.example.example1"
    android:versionCode="1"
    android:versionName="1.0"
    android:sharedUserId="com.sharedID.example">

<manifest xmlns:android="http://schemas.android.
  com/apk/res/android"
    package="com.example.example2"
    android:versionCode="1"
    android:versionName="1.0"
    android:sharedUserId="com.sharedID.example">
```

The following screenshot is displayed when these two applications are running on the device. Notice that the applications, com.example.example1 and com.example. example2, now have the app ID of **app_113**.

```
⊙ ○ ○                                      platform-tools — ddms — adb — 127×9
app_20    1142  80    117196 23776 ffffffff 00000000 S com.google.android.apps.maps
app_14    1231  80    100956 21184 ffffffff 00000000 S com.android.packageinstaller
shell     1369  90    788    356   c007e6cc afd0c4dc S /system/bin/sh
root      1457  2     0      0     ffffffff 00000000 S flush-179:0
app_113   1552  80    114744 20008 ffffffff 00000000 S com.example.example2
root      1563  2     0      0     ffffffff 00000000 S flush-31:0
app_113   1569  80    102356 19920 ffffffff 00000000 S com.example.example1
shell     1582  1369  940    332   00000000 afd0b58c R ps
$
```

You will notice that the shared UID follows a certain format akin to a package name. Any other naming convention will result in an error such as an installation error: `INSTALL_PARSE_FAILED_BAD_SHARED_USER_ID`.

> All applications that share the same UID should have the same certificate.

External storage

Starting with API Level 8, Android provides support to store Android applications (APK files) on external devices, such as an SD card. This helps to free up internal phone memory. Once the APK is moved to external storage, the only memory taken up by the app is the private data of the application stored on internal memory. It is important to note that even for the SD card resident APKs, the **DEX (Dalvik Executable)** files, private data directories, and native shared libraries remain on the internal storage.

Adding an optional attribute in the manifest file enables this feature. The application info screen for such an application either has a move to the SD card or move to a phone button depending on the current storage location of APK. The user then has an option to move the APK file accordingly.

If the external device is un-mounted or the USB mode is set to `Mass Storage` (where the device is used as a disk drive), all the running activities and services hosted on that external device are immediately killed. A detail of the external storage and its security analysis is done in *Chapter 7, Securing Application Data*. In this section, we will discuss how to specify the external storage preference in the policy file.

The feature to enable storing APK on the external devices is enabled by adding the optional attribute `android:installLocation` in the application's manifest file in the `<manifest>` element. The attribute `android:installLocation` can have the following three values:

- `InternalOnly`: The Android system will install the application on the internal storage only. In case of insufficient internal memory, storage errors are returned.

- `PreferExternal`: The Android system will try to install the application on the external storage. In case there is not enough external storage, the application will be installed on the internal storage. The user will have the ability to move the app from external to internal storage and vice versa as desired.

- auto: This option lets the Android system decide the best install location for the application. The default system policy is to install the application on internal storage first. If the system is running low on internal memory, the application is then installed on the external storage. The user will have the ability to move the application from external to internal storage and vice versa as desired.

For example, if android:installLocation is set to Auto, then on devices running a version of Android less than 2.2, the system will ignore this feature and APK will only be installed on the internal memory. The following is the code snippet from an application's manifest file with this option:

```
<manifest package="com.example.android"
  android:versionCode="10"
  android:versionName="2.7.0"
  android:installLocation="auto"
  xmlns:android="http://schemas.android.
    com/apk/res/android">
```

The following is a screenshot of the application with the manifest file as specified previously. You will notice that **Move to SD card** is enabled in this case:

In another application, where `android:installLocation` is not set, the **Move to SD card** is disabled as shown in the following screenshot:

Setting component visibility

Any of the application components namely, activities, services, providers, and receivers can be made discoverable to the external applications. This section discusses the nuances of such scenarios.

Any Activity or Service can be made private by setting `android:exported=false`. This is also the default value for an Activity. See the following two examples of a private Activity:

```
<activity android:name=".Activity1" android:exported="false" />
<activity android:name=".Activity2" />
```

However, if you add an Intent Filter to the Activity, then the Activity becomes discoverable for the Intent in the Intent Filter. Thus, the Intent Filter should never be relied upon as a security boundary. See the following examples for Intent Filter declaration:

```
<activity
    android:name=".Activity1"
    android:label="@string/app_name" >
    <intent-filter>
      <action android:name="android.intent.action.MAIN" />
      <category android:name="android.intent.category.LAUNCHER" />
    </intent-filter>
</activity>
<activity android:name=".Activity2">
    <intent-filter>
      <action  android:name="com.example.android.
        intent.START_ACTIVITY2" />
    </intent-filter>
</activity>
```

Both activities and services can also be secured by an access permission required by the external component. This is done using the `android:permission` attribute of the component tag.

A Content Provider can be set up for private access by using `android:exported=false`. This is also the default value for a provider. In this case, only an application with the same ID can access the provider. This access can be limited even further by setting the `android:permission` attribute of the provider tag.

A Broadcast Receiver can be made private by using `android:exported=false`. This is the default value of the receiver if it does not contain any Intent Filters. In this case, only the components with the same ID can send a broadcast to the receiver. If the receiver contains Intent Filters then it becomes discoverable and the default value of `android:exported` is `false`.

Debugging

During the development of an application, we usually set the application to be in the debug mode. This lets developers see the verbose logs and can get inside the application to check for errors. This is done in the `<application>` tag by setting `android:debuggable` to `true`. To avoid security leaks, it is very important to set this attribute to `false` before releasing the application.

An example of sensitive information that I have seen in my experience includes usernames and passwords, memory dumps, internal server errors, and even some funny personal notes state of a server and a developer's opinion about a piece of code.

The default value of `android:debuggable` is `false`.

Backup

Starting with API level 8, an application can choose a backup agent to back up the device to the cloud or server. This can be set up in the manifest file in the `<application>` tag by setting `android:allowBackup` to `true` and then setting `android:backupAgent` to a class name. The default value of `android:allowBackup` is set to `true` and the application can set it to `false` if it wants to opt out of the backup. There is no default value for `android:backupAgent` and a class name should be specified.

The security implications of such a backup are debatable as services used to back up the data are different and sensitive data, such as usernames and passwords can be compromised.

Putting it all together

The following example puts all the learning we have done so far to analyze `AndroidManifest.xml` provided with an Android SDK sample for `RandomMusicPlayer`.

The manifest file specifies that this is version 1 of the application `com.example.android.musicplayer`. It runs on SDK 14 but supports backwards up to SDK 7. The application uses two permissions namely, `android.permission.INTERNET` and `android.permission.WAKE_LOCK`. The application has one Activity that is the entry point for the application called `MainActivity`, one Service called `MusicService`, and one receiver called `MusicIntentReceiver`.

`MusicService` has defined custom actions called `PLAY`, `REWIND`, `PAUSE`, `SKIP`, `STOP`, and `TOGGLE_PLAYBACK`.

The receiver uses the action intent `android.media.AUDIO_BECOMING_NOISY` and `android.media.MEDIA_BUTTON` defined by the Android system.

None of the components are protected with permissions. An example of an `AndroidManifst.xml` file is shown in the following screenshot:

```xml
<manifest xmlns:android="http://schemas.android.com/apk/res/android"
    package="com.example.android.musicplayer"
    android:versionCode="1"
    android:versionName="1.0">

    <uses-sdk android:minSdkVersion="7" android:targetSdkVersion="14" />

    <uses-permission android:name="android.permission.INTERNET" />
    <uses-permission android:name="android.permission.WAKE_LOCK" />

    <application android:icon="@drawable/ic_launcher" android:label="@string/app_title">

        <activity android:name=".MainActivity"
            android:label="@string/app_title"
            android:theme="@android:style/Theme.Black.NoTitleBar">
            <intent-filter>
                <action android:name="android.intent.action.MAIN" />
                <category android:name="android.intent.category.LAUNCHER" />
            </intent-filter>
        </activity>

        <service android:exported="false" android:name=".MusicService">
            <intent-filter>
                <action android:name="com.example.android.musicplayer.action.TOGGLE_PLAYBACK" />
                <action android:name="com.example.android.musicplayer.action.PLAY" />
                <action android:name="com.example.android.musicplayer.action.PAUSE" />
                <action android:name="com.example.android.musicplayer.action.SKIP" />
                <action android:name="com.example.android.musicplayer.action.REWIND" />
                <action android:name="com.example.android.musicplayer.action.STOP" />
            </intent-filter>
            <intent-filter>
                <action android:name="com.example.android.musicplayer.action.URL" />
                <data android:scheme="http" />
            </intent-filter>
        </service>

        <receiver android:name=".MusicIntentReceiver">
            <intent-filter>
                <action android:name="android.media.AUDIO_BECOMING_NOISY" />
            </intent-filter>
            <intent-filter>
                <action android:name="android.intent.action.MEDIA_BUTTON" />
            </intent-filter>
        </receiver>

    </application>
</manifest>
```

Example checklist

In this section, I have tried to put together an example list that I suggest you refer to whenever you are ready to release a version of your application. This is a very general version and you should adapt it according to your own use case and components. When creating a checklist think about issues that relate to the entire application, those that are specific to a component, and issues that might come up by setting the component and application specification together.

Application level

In this section, I have listed some questions that you should be asking yourself as you define the application specific preferences. They may affect how your application is viewed, stored, and perceived by users. Some application level questions that you may like to ask are as follows:

- Do you want to share resources with other applications that you have developed?
    - ° Did you specify the unique user ID?
    - ° Did you define this unique ID for another application either intentionally or unintentionally?

- Does your application require some capabilities such as camera, Bluetooth, and SMS?
    - ° Does your application need all these permissions?
    - ° Is there another permission that is more restrictive than the one you have defined? Remember the principle of least privilege
    - ° Do all the components of your application need this permission or only a few?
    - ° Check the spellings of all the permissions once again. The application may compile and work even if the permission spelling is incorrect.
    - ° If you have defined this permission, is this the correct one that you need?

- At what API level does the application work?
- What is the minimum API level that your application can support?
- Are there any external libraries that your application needs?
- Did you remember to turn off the debug attribute before you release?
- If you are using a backup agent then remember to mention it here
- Did you remember to set a version number? This will help you during application upgrade
- Do you want to set an auto upgrade?
- Did you remember to sign the application with your release key?

- Sometimes setting a particular screen orientation will not allow your application to be visible on certain devices. For example, if your application only supports portrait mode then it might not appear for devices with landscape mode only.

- Where do you want to install the APK?

- Are there any services that might cease to work if the intent is not received in time?

- Do you want some other application level settings, such as the ability of the system to restore components?

- If defining a new permission, think twice if you really want them. Chances are there is already an existing permission that will cover your use case.

Component level

Some component level questions that you will want to think about in the policy are listed here. These are questions that you should be asking yourself for each component:

- Did you define all components?

- If using the third party libraries in your application, did you define all the components that you will use?

- Was there a particular setting that the third party library expects from your application?

- Do you want this component to be visible to other applications?

- Do you need to add some Intent Filters?

- If the component is not supposed to be visible, did you add Intent Filters? Remember as soon as you add Intent Filters, your component becomes visible.

- Do other components require some special permission to trigger this component?

- Verify the spelling of the permission name.

- Does your application require some capabilities such as camera, Bluetooth, and SMS?

Summary

In this chapter, we learned how to define an applications policy file. The manifest file is the most important artifact of an application and should be defined with utmost care. This manifest file declares the permissions requested by an application and permissions that the external applications need to access its components. With the policy file we also define the storage location of the out APK and the minimum SDK against which the out application will run. The policy file exposes components that are not sensitive to the application. At the end of this chapter we discussed some sample issues that a developer should be aware of when writing a manifest file.

This chapter concludes the first section of the book where we learn about an Android application structure. Let's move to the next section of this book that focuses on secure storage of user data.

5
Respect Your Users

Now that we have a clear understanding of the Android platform and application security framework and components, let's dive into the most challenging aspect of security: data protection. As I stated before, your credibility as an application developer depends on how securely you handle your users' data. Thus the name of this chapter: *Respect Your Users*!

This chapter forms the basis for understanding the importance and significance of securing user data. The chapter starts off with a discussion of benchmarks for assessing the security of the data and the CIA triad. Next, we take an example of our bookstore application and run it through the asset, threat, and attack scenarios. We talk about the mobile ecosystem and how different components of this ecosystem affect the security of user data. We will close with a review of the **Digital Rights Management (DRM)** framework for Android.

Principles of data security

This section discusses the three principles of data security, namely confidentiality, integrity, and availability, usually called the **CIA**. Any piece of data stored on the device or server should meet these three attributes for security. Understanding these benchmarks will help us evaluate how secure our data storage solution is. All these three principles are usually expressed as a CIA triad.

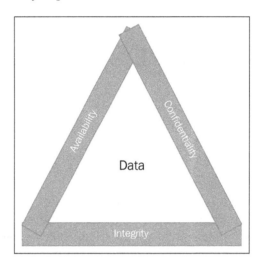

Confidentiality

Confidentiality is the first pillar of security and it focuses on privacy of data. This principle ensures that private data stays away from prying eyes and is only available to users with appropriate access rights. For example, the private data of an Android application should only be accessible to that application's components or other components with appropriate permission (in case the data is protected using permissions). The Linux operating system sandbox and permissions enforces this confidentiality. In another case, an encrypted file containing sensitive data may exist on the SD card. Even if the device or SD card has been compromised, this information will not be leaked out. This kind of confidentiality is enforced using cryptography. Another example of confidentiality is when a device locks itself after a certain period of inactivity and needs user credentials to be unlocked. Note that the Linux kernel does not support filesystem encryption by default, thus encrypting sensitive data before you store it is vital for security.

Integrity

Data integrity ensures that data is not altered or modified in transit or at rest either deliberately or by accident. As an example, inappropriate writing into a database table can cause accidental integrity issues. Therefore, it is always recommended to use built-in synchronization methods to enforce data integrity, unless you know your stuff really well. An example of intentional data integrity breach can be caused during transit where an application is communicating with the server. A man in the middle can listen to the data and alter it as it travels. To mitigate this kind of fraud, it is always recommended to encrypt data and use the **Secure Socket Layer** (SSL) protocol when communicating with the server. For additional security, a checksum can be used. SSL also requires CA's chain of certificate validation that is rarely used in Android applications.

Availability

Data availability ensures that data is available when it is needed. I'd like to add to it and say that data is available when needed by users who have proper rights to access it. This is really important as in the name of availability an application should not let unauthorized users access sensitive information.

Identifying assets, threats, and attacks

There is nothing like absolute security. When we talk about data security, we need to identify what is it that we are protecting and from whom. The following three questions can help us map our approach:

1. *What are we trying to protect?* From an Android application perspective, are we trying to protect the username and password of the user, or the coupon code and credit card number that a user might enter to make a purchase through your application, or rights protected song or picture that the user purchased using your app? By answering this question we can nail down our assets.

2. *Who are we trying to protect the asset from?* In other words, what is our threat? Are we trying to protect user data from other applications on the system, or are we trying to protect this information from other apps that you have developed? Do we want to protect our asset even if the device is stolen?

3. *What is the attack?* Answering this question helps identify vulnerabilities in our application. We get in the mind of the hacker and think how to exploit holes in our application.

Answering the preceding three questions will help us to determine the value of our asset and how much time and energy we are willing to spend on protecting these assets. Let us try to answer the preceding questions with an example application. We go back to our bookstore application where a user can browse through books from the catalog, add books to a wish list, and order books to be shipped to the user. Our application remembers basic information about the user such as the last author and category that the user browsed and the language and username, so that when the user logs in, the app makes certain suggestions and the user feels at home. Our application also provides the user with the store credit card number, mailing address, and name for easy checkout when the user is ready to pay for the book.

Let us try to answer the first question: what are we trying to protect? In the preceding example, our assets are:

- Name
- Credit card number
- Mailing address
- Last author searched
- Last language searched
- Last category searched
- Username
- Password
- Wish list of books

The following figure illustrates different sensitive data artifacts in our example:

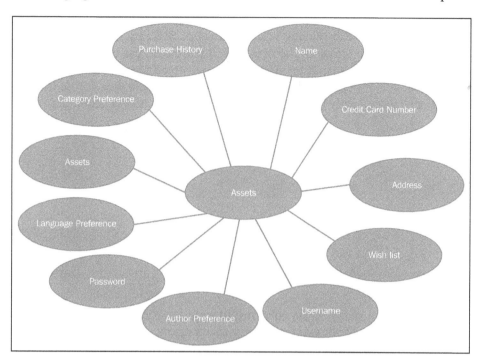

Note that not all of these assets are equally important to protect. A storage mechanism has to be decided upon based on how sensitive a piece of information is. For example, credit card numbers and passwords (if they are being stored on the device) need to be protected ferociously. You can encrypt such information and also store a hash of such information instead of storing this information in raw form. You will encrypt the information in transit and also use SSL protocol for secure communication. Loss of user preferences such as language, author, and category do not possess significant risk. Even if this information is lost, a user can set it up again.

The preceding analysis also triggers the debate about thick and thin clients with respect to PII. Thick clients store a lot of information on the device itself. So an application will end up storing PII on the device as well. Thin clients rely on the backend servers for all the heavy lifting. They store minimal information on the device. This is a good approach as a device can be lost or stolen and then the risk is compromised user data.

Next, we figure out the attack scenarios. Some example scenarios are discussed as follows.

Let us imagine that the user installs a malicious app. This app now tries to steal user information in different ways. In the first case, it tries to access different database tables and extract user information. This is a case of stealing private information. If the database table is protected by permission, we are in a safe position. If the Content Provider checks the identity of the component, we are in an even safer situation.

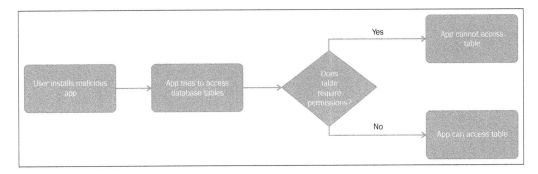

In another scenario, a rogue application might send broadcast messages with bad data that the receiving application might try to act upon, or the malicious app might try to launch other application's components, with ill-formed data that might cause the other application to crash. It is thus important to check the identity of the calling application and vet the input data before acting upon it.

The important lessons from this attack scenario are as follows:

- Never expose a component unless it is absolutely required. Keeping a component private is our first line of defense.

- If we expose a component, we make sure that we protect it with permissions. This is a good place to decide whether we want to expose it to the entire system or just other applications created by you. If the use case is to share components among applications written by the same author, we can define custom permissions.

- Reduce the attack surface by specifying some Intent Filters.

- Always remember to check for input data before acting upon it. If the data is not in the desired format or form, there should be a plan to exit the situation gracefully. Displaying an error message to the user can be an option in this case.

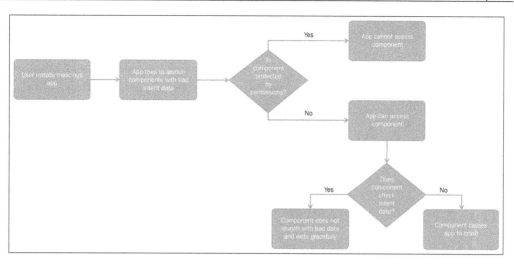

Other scenarios could include a malicious application that listens to data exchange from a device connected to a rogue Wi-Fi. This application can intercept the information, modify it, pretend to be the server that the user is connecting to, or completely block the flow of data. All these scenarios are security risks.

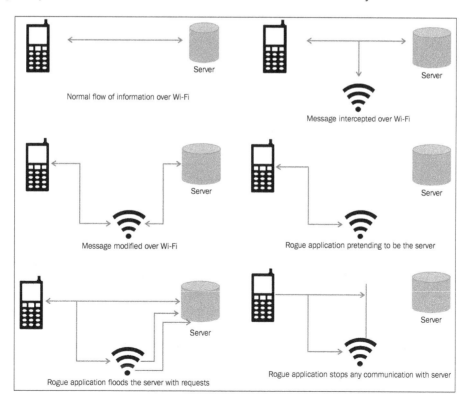

Yet another example is when the malicious application changes the data stored on the device. The user may not even be aware of the fact that this information has been altered. Let us assume that our application was localized in different languages and the user sets the preferred language. In the following scenario, the user's preferred language is changed from English to, say, Japanese. The next time the user logs in, the application opens up in Japanese. In our case, the security risk is not big and it is an annoyance to the user but this example proves the point that information modification is another security risk.

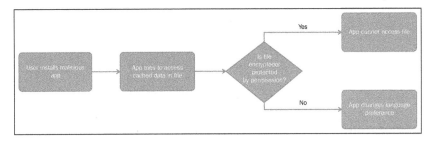

Finally, we need to access the loss in case a security breach happens and our plan of action. If private information such as credit card information, passwords, and social security numbers are stolen, it is a serious security risk. A plan to inform the users in case of a security breech has to be thought through. The user preferences and wish lists are inappropriately accessed; it might cause the user annoyance, but might not be such a privacy risk.

What and where to store

The previous analysis leads us to two important decisions that an application developer has to think upon.

First, an application developer has to decide the pieces of information he/she wants to collect from users. Just as there is a principle of least privilege so there is the principle of least storage. The principle of least storage results in minimizing risk and liability. An application developer should always try to offload the storage of **Personally Identifiable Information (PII)**. In our previous example, the application might not like to store credit card details, billing addresses, and other information related to payments. Payments are a tricky domain and companies such as PayPal can assist the user with the checkout process. Also any application that processes credit card numbers is recommended to follow the **PCI (Payment Card Industry)** standard. This standard lists requirements that such an application and server must comply with. My suggestion is to off-load such action to services that know these things best.

The second important decision to ponder upon is where to store the user data. In the distributed data storage environment of today, a developer has many options of storage such as on device, on a server, on the cloud, or a third party application. A mobile device should never be considered as a secure storage location, partly because it can be easily stolen or lost and also since most devices do not possess the sophisticated security mechanisms such as secure element and dual boot that desktops and laptops possess. Passwords, cryptographic keys, large content files, PII, and other sensitive data should be stored on the backend servers. Again, it is important to firewall these servers.

We will come back to this example in *Chapter 7, Security Application Data*, where based on the preceding analysis we decide the appropriate storage options and protection mechanisms.

End-to-end security

About a decade ago, we stored our music on tapes and disks; our pictures resided in albums, and we used a phone for emergency purposes only. Fast forward to today; more and more of our lives are going digital. Our friends, families, likes, dislikes, pictures, contact lists, and even our purchase histories and credit card numbers are going digital. Imagine the scenario where a user loses their phone. Besides the monetary value of the device and emotional value associated with content stored on it, the biggest risk is the compromise of the user's personal information that is stored on the device. This information could include PII, which can identify an individual such as name, social security number, date of birth, and mother's maiden name. It could also include access to passwords, contact list, and SMS data. This risk lurks even if the user who is in possession of the device and the device is compromised due to malware.

The mobile ecosystem

As illustrated in the following figure, there are different artifacts in the mobile ecosystem such as device, networks, applications that a user installs on devices, OEMs (Original Equipment Manufacturers), and other services that a consumer's device interacts with.

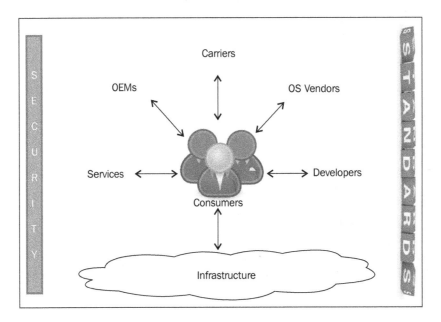

Let us look at these components a little deeper.

- **Consumers**: The entire ecosystem revolves around the consumers and how the consumers interact with different pieces of ecosystem.

- **Device manufacturers**: Also called OEMs, these are companies that produce the hardware for the device. Companies such as HTC, Motorola, Samsung, and LG all design and manufacture Android devices. Besides the size and style of the device, each device manufacturer puts in their **systems on chip (SOC)**, device drivers, and firmware that affect how applications work on different devices. If you have tested your application on different devices, you can easily notice these differences. Any security flaw at the hardware level affects all devices that use this hardware. Hardware flaws are also difficult to patch.

- **Operating system vendors**: Android is an open source operating system and manufacturers are at liberty to modify it or use their own pieces of software. For example, a device manufacturer may decide to use a different WebKit engine, music player, or screen instead of the one that comes bundled with the Android stack. This will cause applications to behave and look different on different devices. A security flaw in these such proprietary packages may cause your application to be compromised. All devices running a particular version of the operating system are affected by a defect. Defects at the software level can usually be patched and users are recommended to keep their software updated at all times.

- **Carriers**: AT&T, Sprint, Verizon, Orange, and Vodafone are all carriers that provide the infrastructure that makes mobile devices truly mobile. They provide the data and voice plans for our devices. They also work with device manufacturers (who are in most cases the operating system vendors as well) to bundle their custom applications in system image. They might also work with OEMs to adapt security rules to suit their needs. For example, they may request an OEM to directly load and install applications without asking the user for consent or showing permission requests.

- **Services**: These are services that a device interacts with such as cloud services for backup. In most cases, the user installs a client that interacts with the backend. Other services could be payment services such as PayPal, mailing services such as Gmail, and social networking services such as Facebook and Twitter. Most of these services are provided to the user as third party applications.

- **Application developers**: This is the category of individual application developers or small groups of developers that put their applications on the app stores such as Google Play and Amazon appstore. Examples of such applications include utility applications, games, content consumption applications. Most of the audience of this book belongs to this category.

- **Infrastructure**: These are the technologies and protocols that are the backbone of mobile infrastructure. These include **CDMA (Code Division Multiple Access)**, **GSM (Global System for Mobile)**, **WiMAX (Worldwide Interoperability for Microwave Access)**, **WAP (Wireless Application Protocol)**, and proximity technologies such as NFC, RFID, and Bluetooth. Security flaws in these technologies can render our applications susceptible to attacks.

- **Standards and security**: These are two pieces of the mobile ecosystem that are still being evolved as we write this book.

As you will have noticed, there are many actors in the mobile ecosystem, thus increasing the risks and threat surface. Moreover, not all major players in the mobile world work together and in some cases work against each other resulting in complex attack models. Also, manufacturers produce devices for a targeted market. It is thus a complex landscape with moving and evolving parts. Looking at security from an end-to-end perspective, it is not hard to realize that the only power application developers have is on the applications that they create. Any other flaw in the device or operating system can cause a security breach as well. For example, a flaw in the operating system can cause escalation of privileges and let an app act as root. In this case, this root application can access all information on the device. All applications will be compromised but if the developer used good security standards, their liability is minimal.

 The only power that application developers have is on their own application. Any malicious user can exploit a weakness in device hardware, operating system, or carrier application and gain access to user data.

For example, our bookstore app talks to the database, sends information to the server and caches some data as well. All these cases need to be protected. If the device is using some kind of proximity technology such as **Near Field Communication (NFC)**, Bluetooth, or **Radio Frequency Identification (RFID)**, to exchange data, it is important to understand the security risks and new attached scenarios associated with these technologies.

Chapter 6, Your Tools – Crypto APIs, discusses cryptographic algorithms that can be used to secure data in transit.

Three states of data

Let us look at the information flow in a typical mobile application. Consider the bookstore application once again. In our bookstore application, a user can browse through books from the catalog, add books to a wish list, and order books to be shipped to the user. Our applications remember basic information about the users such as the last author and category that the user browsed, and the language and username. The user's credit card number, mailing address, and name are also stored for easy checkout.

The following figure shows one possible scenario. The bookstore application uses a SQLite database and flat files on the Android device as the cache. The application stores account details, book catalogues, and the wish list on the external server and connects to backend servers using Wi-Fi.

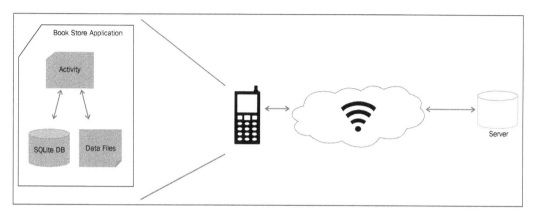

At any given point, data can be either resting at a location, be in transit from one node to the other, or is in the process of being worked upon. We call these three states of data data at rest, data in transit, and data in use. Let's look at these three more closely:

1. **Data at rest**: This is data that is stored on some kind of storage media such as SD cards, device memory, backend servers, and databases. This data is in an inactive state. In the preceding example, data residing in flat files, SQLite database tables, and on the backend server, is all considered as data at rest.

2. **Data in use**: Data that is currently being processed is called data in use. Examples of this kind of data includes data that is being accessed from database tables, data sent to application components with intents, and a file that is currently being written to or read from.

3. **Data in transit**: When data is being transferred from one node to another it is termed as data in transit. Data being transferred to the application from the database in response to a query is an example of data in transit.

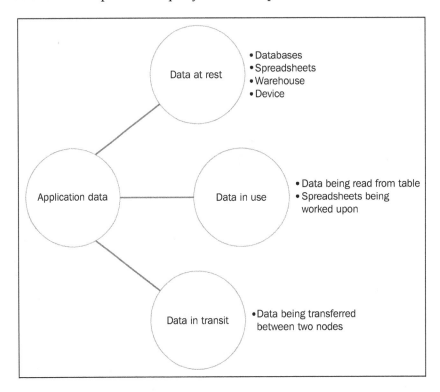

When dealing with data and thinking about end-to-end security, it is important to protect data in all three states.

Digital rights management

Digital rights management (**DRM**) is access control technology for digital content such as music, ebooks, applications, video, and movies. Access control is based on a rights object associated with content. This rights object contains rules that limit the use, distribution, and duplication of content. DRM schemes such as OMA DRM v1 and OMA DRM v2 are developed by **Open Mobile Alliance** (**OMA**) but many device manufacturers have their proprietary DRM schemes as well.

A DRM system contains the following components:

- **Content server**: This is the server from where the device pulls the media content.

- **Rights server**: The server from where the device pulls in the rights object. The rights object is usually an XML file with permissions and constraints associated with the content.

- **DRM agent**: The agent lives within the device and is the trusted body that associates content and rights and enforcement of rights and permissions on content.

- **Storage device**: This is the device where the content and the rights objects are stored. It could be a phone or a tablet, or external storage such as an SD card or even cloud storage.

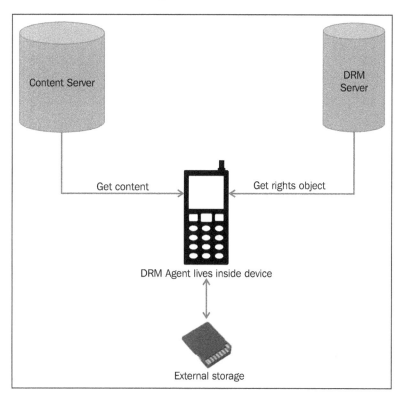

You can read the complete specifications about OMA DRM at www. openmobilealliance.org. OMA DRM 1.0 supports models such as forward locking of content (content cannot be forwarded to another device), combined delivery (content and rights objects are delivered together), and separate delivery (content and rights objects are pulled separately from different servers). Security of OMA DRM v2.0 is based on PKI, which is significantly more secure. Manufacturers can pick and choose the DRM scheme they want to support on their device. They can also implement or modify the DRM scheme accordingly.

Android started supporting DRM in API 11. Support for DRM in Android is open so that manufacturers can choose their own DRM agents. This is achieved by implementing the DRM framework in two architectural layers. The Android developer website (developer.android.com) shows it diagrammatically as follows:

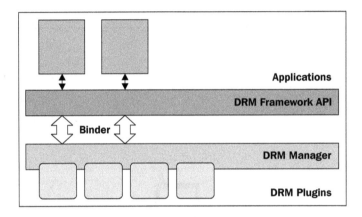

The DRM manager implements the DRM framework and is of interest to device manufacturers who integrate their DRM agents of choice with this framework as plugins. The framework layer abstracts all the complexity of the DRM manager and exposes to the developers a unified set of APIs to work with. These APIs run in the Dalvik VM with the rest of the application code.

All DRM APIs are present in the android.drm package. This package has classes and interfaces to get rights information, associating content with rights, querying for DRM plugin and MIME types. The DrmManager class provides each DrmManagerClient with a unique ID to operate with it.

The application first needs to *find out the DRM plugins* available on the device. This can be accomplished using the DrmManagerClient class.

```
DrmManagerClient mDrmManagerClient = new
DrmManagerClient(getContext());
String[] plugins = mDrmManagerClient.getAvailableDrmEngines();
```

The next step is to *register with the DRM server* and download the rights object.

```
DrmManagerClient mDrmManagerClient = new DrmManagerClient(context);
DrmInfoRequest infoRequest = new DrmInfoRequest(DrmInfoRequest.TYPE_
RIGHTS_ACQUISITION_INFO, MIME);
mDrmManagerClient.acquireDrmInfo(infoRequest);
```

The third step is to *extract license information* from the rights object. This is done using the getConstraints method of DrmManager.

```
ContentValues constraintsValues = mDrmManager.getConstraints(String
path, int action);
ContentValues constraintsValues = mDrmManager.getConstraints(Uri uri,
int action);
```

Now, we need to *associate content with the rights object*. This is done by specifying the content path and the rights path in the saveRights method of DrmManager. Once this association is done, the DRM agent will continue to enforce permissions on content without user interference.

```
int status = mDrmManager.saveRights(DrmRights  drmRights, String
rightsPath, String  contentPath);
```

The android.drm package provides some other utility functions as well. Check this package (https://developer.android.com/reference/android/drm/package-summary.html) for all the functions available there.

Summary

This chapter covered the basics of data security. We discussed the three core principles of data security, namely confidentiality, integrity, and availability. We walked through a sample application scenario and tried to chart out our assets, threats, and attack scenarios. We tried to assess the cost associated with a security breach. Our data storage options and the amount of time, effort, and money that we plan to spend on securing the data will depend on this analysis. We also reflected on the mobile ecosystem as a whole and what end-to-end security means in a mobile context. It is not hard to realize that we only control the applications that we write. We closed this chapter with the review of Android's DRM framework and available capability. With all this knowledge about data security, let's march to the next chapter and learn the different tools that an application developer can use to protect their users' data.

6
Your Tools – Crypto APIs

To respect user privacy, applications that process sensitive data need to protect this data from prying eyes. Although the Android stack provides layered security architecture with security built in the operating system itself, it is relatively easy to gain root access on the device, thereby compromising the data stored on the device. It is thus important for application developers to be aware of the tools that they can use to securely store data. On the same note, it is important for them to understand how to properly transmit data.

The Android stack provides tools that developers can use to perform tasks such as encryption and decryption, hashing, generating random numbers, and message authentication codes. These tools are the cryptographic APIs provided by various packages in the stack. The `javax.crypto` package provides capabilities to encrypt and decrypt messages, and generate message authentication codes and key agreement APIs. Random number generation is provided as a utility by the `java.util.Random` class, and the `java.security` package provides APIs for hashing, key generation, and certificate management.

In this chapter, we will discuss crypto APIs provided by the Android stack and available to application developers to protect sensitive information. We begin the basic terminology used in cryptography, followed by information on how to find out the security providers available. Next, we will discuss random number generation followed by hashing functions. Asymmetric and symmetric key cryptography and different cipher modes are discussed next followed by message authentication codes.

Terminology

Let us start off with an understanding of some terms that are used in cryptography. As we progress through the chapter, these terms will be repeatedly used, so it is important to become familiar with them before we proceed.

- Cryptography: Cryptography is the study and practice of secure communication in an insecure environment and in the presence of adversaries. As our lives become more digitalized and connected, cryptography has gained increasing importance. Cryptography is practised in the form of algorithms and protocols that are designed using mathematical formulae and problems that are computationally hard.

- Plaintext: Also called cleartext, plaintext is the message that a sender wants to transmit and that needs to be kept a secret. If Alice wants to send a message "Hello World" to Bob, then "Hello World" is the plaintext.

- Ciphertext: Also called codetext, this is the encoded or encrypted message of plaintext that is sent to the receiver. Let us follow the previous example where Alice wants to sends the message "Hello World" to Bob. Alice uses a substitution method where each alphabet is replaced by the next alphabet to form the ciphertext. So, the plaintext "Hello World" is now transformed into "Ifmmp Xpsme". "Ifmmp Xpsme" is the ciphertext that is transmitted to Bob.

- Encryption: Encryption is the process of converting plaintext into ciphertext such that an eavesdropper cannot decipher the message as it is being transmitted or stored, and only the parties that know the code can understand it. In the preceding example the process of converting "Hello World" to "Ifmmp Xpsme" is called encryption.

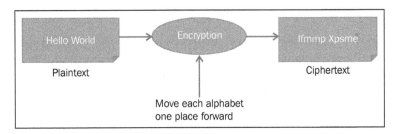

- Decryption: Decryption is the reverse of encryption. It is the process of converting a ciphertext back to a plaintext at the receiving end to retrieve the information. So, the conversion of "Ifmmp Xpsme" back to "Hello World" is called decryption.

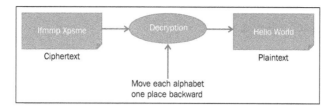

Key: In cryptographic terms, a key is the critical piece of information or mathematical parameter that determines the output of a cryptographic algorithm. In the preceding example, when "Hello World" is converted to "Ifmmp Xpsme", the critical information is adding one to each alphabet and this is the key. During decryption, the critical information is subtracting one from each alphabet and this is the key for decryption.

Cipher: A cipher is the cryptographic algorithm that performs the encryption and decryption of messages. It is also called a cryptographic algorithm. In the preceding example, the cipher is an algorithm of encrypting "Hello World" to "Ifmmp Xpsme" and then at the receiving end, converting "Ifmmp Xpsme" back to "Hello World".

Security providers

The Android stack is customizable as far as the security providers are concerned. This means that the device manufacturers can add their own crypto providers. As an application developer, you are at liberty to use your own security provider as well. Since the Android stack provides only some capabilities of the Bouncy Castle security provider, Spongy Castle is hugely popular. Also, different versions of the Android stack keep updating their crypto capabilities by removing cryptographic algorithms that are way insecure and adding new ones. You may like to check the providers and their complete list of algorithms supported at a given point of time. Also, make sure to test your application on different devices to confirm that the crypto algorithms work as expected.

The following code snippet shows how to obtain a list of crypto providers by using the `java.security.Providers` method:

```
for (Provider provider: Security.getProviders()) {
    System.out.println(provider.getName());
}
```

```
● ○ ○                                    platform-tools — adb — 140×7
D/dalvikvm(   37): GC_EXPLICIT freed <1K, 4% free 9032K/9347K, paused 3ms+13ms
I/System.out(  590): AndroidOpenSSL
I/System.out(  590): DRLCertFactory
I/System.out(  590): BC
I/System.out(  590): Crypto
I/System.out(  590): HarmonyJSSE
```

Now to get the detailed information about each provider, let's enhance the function to log more details, as follows:

```
for (Provider provider: Security.getProviders()) {
    System.out.println(provider.getName());
    for (String key: provider.stringPropertyNames()) {
      System.out.println("\t" + key +
        "\t" + provider.getProperty(key));
    }
}
```

The following screenshot shows the details about some security providers:

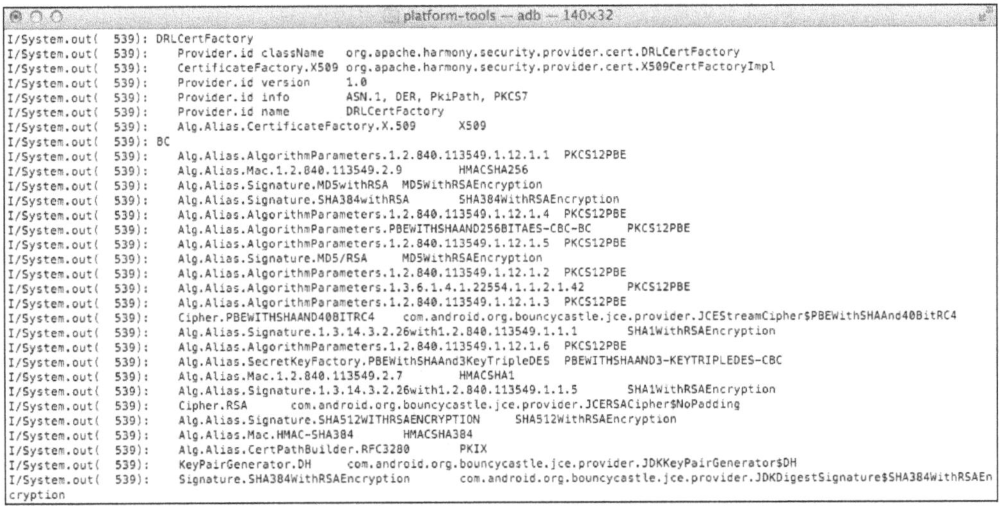

Always use well-known, industry standard cryptographic algorithms. Writing a crypto routine sounds fun and easy, but it is much harder that it seems. Industry standard algorithms like we will study in the following section have been developed by cryptographic experts and thoroughly tested. If any weakness is found in such algorithms, then that is made aware to the public, and developers can update their code with stronger crypto algorithms.

Random number generation

Generating a random number is one of the most important tasks in cryptography. A random number acts as a seed for other crypto functions, such as encryption and generating message authentication codes. It is hard to simulate the generation of true random numbers as it comes from unpredictable acts of nature. Computer systems generate pseudo random numbers which means that they are not truly random but appear random.

There are two approaches to compute generated random numbers: **Pseudo Random Number Generators (PRNG)** and **True Random Number Generators (TRNG)**. PRNGs are generated by an algorithm based on some mathematical formula. TRNGs are based on system characteristics, such as **CPU (Central Processing Unit)** cycles, clock, noise, and keystrokes. Dr. Mads Haahr, professor at Trinity College runs www.random.org and this is a very interesting stop for anyone who is interested in randomness. Check it out!

Use cases of random numbers include gaming applications such as those where users roll a dice, gambling applications, music applications that play songs randomly, and as a seed for crypto operations like hashing, encryption, or key generation. Not all use cases require strong randomness. A music player playing tracks randomly does not require as strong randomness as a key generation algorithm might require.

Android provides the ability to generate random numbers using the java.util. Random class of the java.util package. The class provides methods to generate one or more arrays of random double, byte, float, int, or long. This class is thread-safe.

The following code snippet shows an example of how to generate a random number within a range of 1 to 100.

```
int min = 1;
int max = 100;

public int getRandom(int min, int max) {
  Random random = new Random();
  int num = random.nextInt(max - min + 1) + min;
  return num;
}
```

A random number can be generated using a seed as well. However, since the Android stack has a pseudo random number generator that seeds itself with an initial state quite unpredictable, the seed actually makes the random number more predictable.

Hashing functions

Hashing functions are algorithms that work on data of an arbitrary length to produce a fixed length output. Given the same input, the output is always the same and for different input values, the output is always different. These functions are one way, which means that the reverse operation on data is not possible.

In mathematical terms, a one-way hash function can be defined as follows:

Given a message M, and a one-way hash function H, it is easy to compute x such that $H(M) = x$. But given x and H, it is infeasible to get the message M. This can be shown mathematically as follows:

$H(M) = x$

$H(x) \neq M$

Another property of hash functions is low collision probability. This means that given a message M, it is hard to find another message M, such that:

$H(M) \neq H(M')$

One-way hash functions can be used for various applications. They are used to create a fixed size output for a variable length string. Using a hash, a value can be securely stored as given by the hash; it is unfeasible to retrieve the original message. For example, instead of storing passwords, the hash of passwords is stored in the table. Since the hash value is always the same for a given message, entering the correct password will lead to the generation of the same hash value. They are used as a checksum to ensure that the message is not altered in transit.

The most popular hash functions that are used today are the **MD5 (Message Digest Algorithm)** and **SHA (Secure Hash Algorithm)** family of hash functions. All these hash functions are different in strength and collision probability, and you should use the one best suited for your application. Usually, using SHA-256 is a good choice. Many applications still use MD5 and SHA-1 but these are now considered secure enough. For applications that require very high-level of security, stronger hash functions, such as SHA-3 should be considered. The following table summarizes the length of the output for some common hashing functions:

Hash algorithm	Block length (in bits)	Output length (in bits)
MD5	512	128
SHA-1	512	160
SHA-256	512	256
SHA-512	1024	512

The following image from Wikipedia shows how small changes in input change the output completely. The hash function in this case is SHA-1:

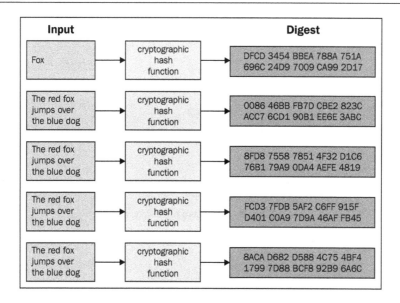

Hashing functionality is provided by the `java.security.MessageDigest` class of the `java.security` package. The following code snippet shows how to use this class to create a SHA-256 hash on string `s`. The method `update` updates the digest with bytes and the method `digest` creates the final digest.

```
final MessageDigest digest = MessageDigest.getInstance("SHA-256");
digest.update(s.getBytes());
byte messageDigest[] = digest.digest();
```

Public key cryptography

Public key cryptography is a cryptographic system that uses two keys: one for encryption and one for decryption. One of the keys is made public and the other is kept private.

Public key cryptography is most commonly used to target two use cases. One for confidentiality and the other is for authentication. In case of confidentiality, the sender encrypts the message using the receiver's public key and sends it over. Since the private key is in possession of the receiver, the receiver uses the private key to decrypt the message.

In the case of authentication to serve as a digital signature, a sender uses their private key to encrypt the message (in most use cases, it is the hash of the message that is encrypted and not the entire message) and makes it available. Anyone with a public key can access it and be certain that the message comes from the sender.

Both the use cases are shown in the following screenshot:

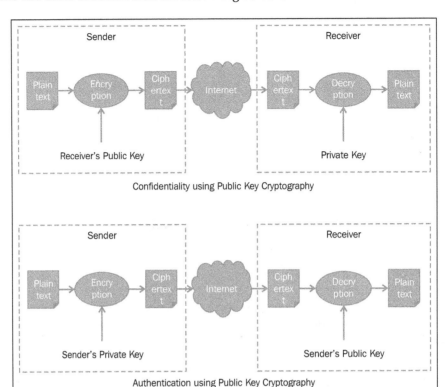

Confidentiality using Public Key Cryptography

Authentication using Public Key Cryptography

In the following section, we discuss two common public key cryptography algorithms: RSA for encryption and authentication, and Diffie-Hellman for key exchange.

RSA

Named after its inventors Ron Rivest, Adi Shamir, and Leonard Adleman, RSA is an algorithm based on the public key cryptography. The security of RSA is based on factoring two large primes. The algorithm itself is not a secret and neither is the public key. Only the primes are secret.

The length of the RSA key used can be 512, 1024, 2048, or 4096 bits based on the strength required. Currently the 2048 bit key is considered strong. RSA is very slow, so its use to encrypt large data sets should be avoided. It is important to note that the length of the message that can be encrypted with RSA cannot exceed the length of modulus (length of the product of the two primes). Since RSA is inherently slow, the usual approach is to encrypt the plaintext with a symmetric key and then encrypt the key with RSA.

RSA can be used both for confidentiality and authentication using the digital signature. There are three main operations when using RSA which are discussed, as follows:

Key generation

The first step in implementing RSA is to generate the keys. In Android, this can be done by using the `java.security.KeyPairGenerator` class. The following code snippet shows how to generate a 2048 bit key pair:

```
KeyPairGenerator keyGen = KeyPairGenerator.getInstance("RSA");
keyGen.initialize(2048);
KeyPair key = keyGen.generateKeyPair();
```

If the key is already available in raw and the private and public keys need to be extracted from it, then the `java.security.KeyFactory` class can be used to extract the public and private keys from the key specs, as follows:

```
KeyFactory keyFactory = KeyFactory.getInstance("RSA");
keyFactory.generatePublic(keySpecs);
```

Encryption

Both encryption and decryption can be performed by either private or public keys based on the use case. The following code snippet encrypts the data with the public key of the receiver. This example follows from the preceding method where a key pair is generated by using the `java.security.KeyPairGenerator` class. The following example uses the `java.security.Cipher` class to initialize the cipher and perform the operation:

```
private String rsaEncrypt (String plainText) {
    Cipher cipher = Cipher.getInstance("RSA/ECB/PKCS1Padding");
    PublicKey publicKey = key.getPublic();
    cipher.init(Cipher.ENCRYPT_MODE, publicKey);
    byte [] cipherBytes = cipher.doFinal(plainText.getBytes());
    String cipherText = new String(cipherBytes,
        "UTF8").toString();
    return cipherText;
}
```

Decryption

Decryption is the opposite operation of encryption. The following code shows how to decrypt data using the private key. Following from the preceding example, this is the case where the sender encrypts the message using the receiver's public key and then the receiver decrypts it, by using their private key.

```
private String rsaDecrypt (String cipherText) {
    Cipher cipher = Cipher.getInstance("RSA/ECB/PKCS1Padding");
    PrivateKey privateKey = key.getPrivate();
    cipher.init(Cipher.DECRYPT_MODE, privateKey);
    byte [] plainBytes = cipher.doFinal(cipherText.getBytes());
    String plainText = new String(plainBytes, "UTF8").toString();
    return plainText;
}
```

Padding

In the previous examples you will notice that the cipher is initialized with **PKCS1Padding**. Let us talk more about padding. The RSA algorithm has no random components. This means that the same plaintext when encrypted with the same key will result in the same ciphertext. This property can lead to a chosen plaintext attack against the cryptosystem. Before encrypting the plaintext, it is usually padded with random data. **PKCS#1 (Public Key Cryptography Standard)** published by the RSA Laboratories, is used to embed structured random data in plaintext. It was later shown that even PKCS#1 padding is not enough to avoid adaptive chosen plaintext attacks. This is a kind of chosen ciphertext attack in which subsequent ciphers are chosen based on results from the first set of decrypted ciphertexts. To mitigate these kinds of attacks, PKCS#1 v1.5 is recommended. Another kind of padding that can be used is **OAEP (Optical Asymmetric Encryption Padding)**.

In the example you will also notice **CBC (Cipher Block Chaining)** as the parameter. This mode is discussed later in this chapter, under the *Block Cipher Modes* section.

The Diffie-Hellman algorithm

Published by Whitefield Diffie and Martin Hellman in 1976, Diffie-Hellman is the most popular key exchange algorithm. The beauty of this algorithm is that two parties can independently generate a secret key over an insecure channel without exchanging the secret key. This secret key can then be used in symmetric encryption.

The Diffie-Hellman algorithm does not authenticate the two parties. It is thus susceptible to the man-in-the-middle attack where an eavesdropper sits in the middle and communicates with the two parties posing as the other party. The following illustration from Wikipedia explains the concept of Diffie-Hellman beautifully using two parties: Alice and Bob:

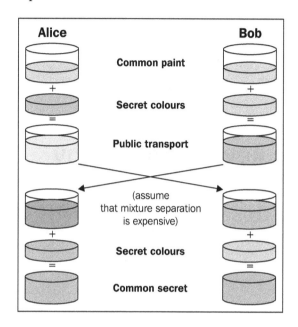

The following code sample shows an example implementation of generating key pairs. The `java.security.KeyPairGenerator` class is used to generate the key pair based on the DH parameters. Next, the `javax.crypto` class is used to generate the key agreement:

```
// DH params
BigInteger g = new BigInteger("0123456789", 16);
BigInteger p = new BigInteger("0123456789", 16);
DHParameterSpec dhParams = new DHParameterSpec(p, g);

// Generate Key pair
KeyPairGenerator keyGen = KeyPairGenerator.getInstance("DH");
keyGen.initialize(dhParams, new SecureRandom());

// Generate individual keys
```

```
KeyAgreement aKeyAgree = KeyAgreement.getInstance("DH");
KeyPair aPair = keyGen.generateKeyPair();
aKeyAgree.init(aPair.getPrivate());

KeyAgreement bKeyAgree = KeyAgreement.getInstance("DH");
KeyPair bPair = keyGen.generateKeyPair();
bKeyAgree.init(bPair.getPrivate());

// Do the final phase of key agreement using other party's
   public key
aKeyAgree.doPhase(bPair.getPublic(), true);
bKeyAgree.doPhase(aPair.getPublic(), true);
```

Symmetric key cryptography

The symmetric key cryptography is based on a secret key that is the same for both parties. The same key is used for both encryption and decryption. This is a problem compared to the public key cryptography, as it is required to exchange the secret keys securely by some means. If an eavesdropper gets hold of the key, the security of the system is defeated.

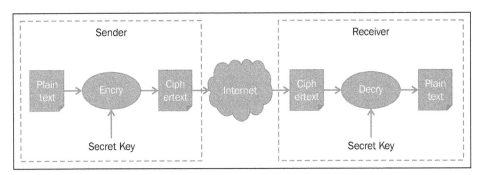

Symmetric key is much faster than a public key and is ideal when encrypting/decrypting large chunks of data. Security of a symmetric key algorithm is based on the length of the key.

Stream cipher

A stream cipher is a type of a symmetric key cryptography where each bit or byte of data is encrypted individually with a stream of random bits called a key stream. Usually, each bit or byte is **XOR**ed (**Exclusive OR**) with a key stream. The length of the key stream is the same as the length of data. The security of a stream cipher depends upon the randomness of the key stream. If the same key stream is used for encrypting multiple data sets, then vulnerability in the algorithm can be identified and exploited. The following image shows a stream cipher in action:

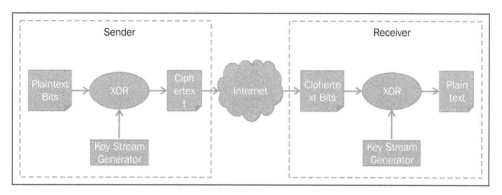

The best use of a stream cipher is where the length of data is variable like in Wi-Fi or an encrypting voice data. They are also easy to implement in hardware. Some examples of algorithms that use stream cipher technique include RC4, A5/1, A5/2, and Helix.

Since the key is as long as the data in question, there are severe key management problems with stream ciphers.

Block cipher

In the case of a block cipher, a block of data is encrypted with a key one at a time. The plaintext is divided into fixed length blocks and each block is encrypted individually. The following figure shows the basic idea of a block cipher. Each plaintext is divided in fixed blocks of data. If the blocks cannot be evenly divided, they are padded with a standard set of bits to make them the desired length. Each block is then encrypted with a key and a fixed length encrypted block is generated.

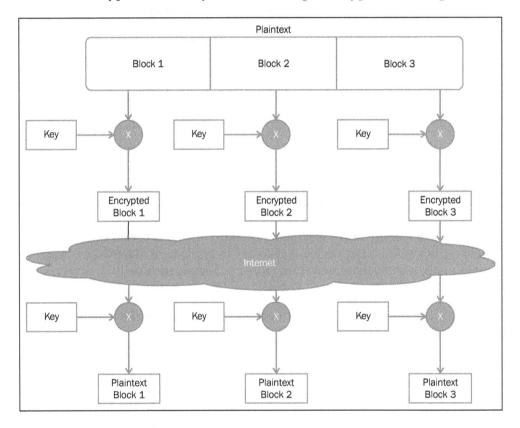

A problem with block ciphers is that if the same block of data is repeated, the output is always the same. Another problem is that if a block is lost in transit, there is no way to identify that a block has been lost. Various block cipher modes have been designed that help solve the previously mentioned problems. Block ciphers are widely used in cryptographic algorithms, such as AES, DES, RC5, and Blowfish.

As plaintext is divided into blocks, it is common that the last block will not have enough bits to fill the block. In such a case, the last block is padded with additional bits to attain the desired length. This process is known as padding.

Block cipher modes

In the block cipher mode, the plaintext is divided into blocks and each block is encrypted with the same key. In the following section some techniques that are used to realize block encryption are discussed. These modes are used both for symmetric encryption and also for asymmetric encryption, such as RSA. In practice though, large chunks of data are rarely encrypted using the asymmetric ciphers, as these tend to be very slow.

Electronic Code Book (ECB)

In the ECB mode, a plaintext is divided into blocks and each block is independently encrypted with the key. This mode can be easily parallelized and is therefore fast. This mode does not conceal patterns in plaintext. So, the same blocks will yield the same ciphertext. Any attacker can modify or steal the plaintext and it will be oblivious to the sender.

The following figure shows how encryption and decryption is realized in the ECB mode:

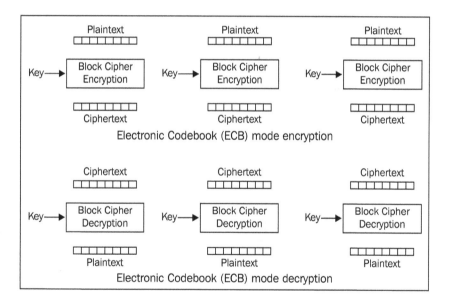

The following code illustrates how to initialize the RSA cipher with the ECB mode:

```
Cipher cipher = Cipher.getInstance("RSA/ECB/PKCS1Padding");
```

Similarly, to initialize an AES symmetric algorithm with ECB, the following code can be used:

```
Cipher cipher = Cipher.getInstance("AES/ECB");
```

Cipher Block Chaining (CBC)

In the CBC mode, each block of plaintext is XORed with the previous ciphertext and then encrypted. This mode takes care of the two shortcomings associated with the ECB mode. XORing the block with the previous plaintext block conceals any patterns in the plaintext. Additionally, except for the first and the last block if any other block is removed or altered, the receiver easily detects it.

The following figure illustrates encryption and decryption of a plaintext block with the CBC mode. Note the use of an **Initialization Vector (IV)** for adding randomness to the first block. IV is a random set of bits that are XORed with the first block:

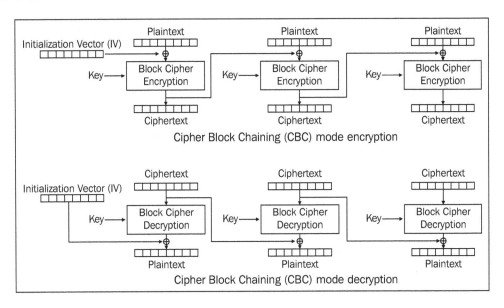

The following code illustrates how to initialize the RSA cipher with the CBC mode:

```
Cipher cipher = Cipher.getInstance("RSA/CBC/PKCS1Padding");
```

Similarly, to initialize an AES symmetric algorithm with CBC, the following code can be used:

```
Cipher cipher = Cipher.getInstance("AES/CBC");
```

Cipher Feedback Chaining (CFB)

In the CFB mode, the previous ciphertext is first encrypted and then XORed with the plaintext to produce the ciphertext. This mode also conceals the plaintext patterns and creates dependency of one plaintext block on the previous block. This enables the tracking and integrity verification of blocks during transmission. Again, note the use of IV for the first block.

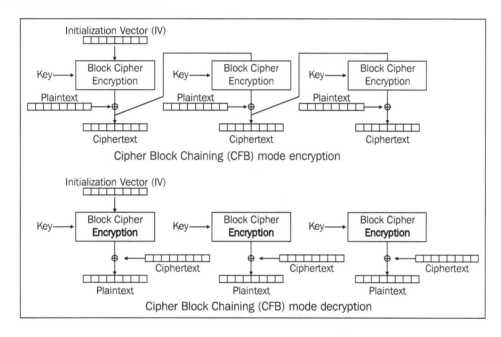

The following code illustrates how to initialize the RSA cipher with the CFB mode:

```
Cipher cipher = Cipher.getInstance("RSA/ECB/PKCS1Padding");
```

Similarly, to initialize an AES symmetric algorithm with CFB, the following code can be used:

```
Cipher cipher = Cipher.getInstance("AES/CFB");
```

Output Feedback Mode (OFB)

OFB is similar to the CFB mode except that the XORed ciphertext acts as a synchronous stream cipher such that an error in one bit affects only one bit and not the entire block. Again, an IV is used to seed the process as follows:

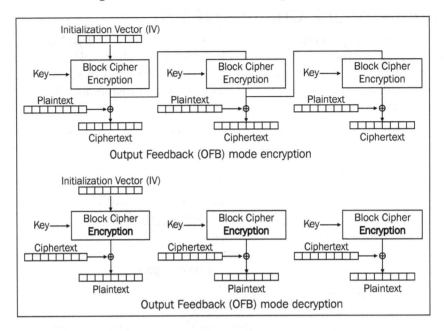

The following code illustrates how to initialize the RSA cipher with the OFB mode:

```
Cipher cipher = Cipher.getInstance("RSA/OFB/PKCS1Padding");
```

Similarly, to initialize an AES symmetric algorithm with OFB, the following code can be used:

```
Cipher cipher = Cipher.getInstance("AES/OFB");
```

Advanced Encryption Standard (AES)

AES is the most popular block symmetric cipher. It is much more secure than the other common block symmetric ciphers, such as DES and DES3. This cipher divides the plaintext into fixed block sizes of 128 bits and the keys could be 128 bit, 192, or 256 bit keys. AES is fast and has low memory requirements. The Android disk encryption also uses AES 128 bit encryption and the master key is encrypted with AES 128 bit encryption as well.

The following code snippet shows how to generate a 128 bit AES key:

```
//Generate individual keys
Cipher cipher = Cipher.getInstance("AES");
KeyGenerator keyGen = KeyGenerator.getInstance("AES");
generator.init(128);
Key secretKey = keyGen.generateKey();
byte[] key = skey.getEncoded();
```

Next, the following code shows how to encrypt a plaintext with an AES key:

```
byte[] plaintext = "plainText".getBytes();
SecretKeySpec skeySpec = new SecretKeySpec(raw, "AES");
Cipher cipher = Cipher.getInstance("AES");
cipher.init(Cipher.ENCRYPT_MODE, skeySpec);
byte[] cipherText = cipher.doFinal(plainText);
```

Following from the preceding example, to decrypt with AES, the following code can be used:

```
SecretKeySpec skeySpec = new SecretKeySpec(raw, "AES");
Cipher cipher = Cipher.getInstance("AES");
cipher.init(Cipher.ENCRYPT_MODE, skeySpec);
byte[] encrypted = cipher.doFinal(cipherText);
```

Message Authentication Codes

A **Message Authentication Code (MAC)** is a tag or checksum that is appended to a message to ascertain its authenticity and integrity. Authentication is provided by the possession of a secret key, and verifying accidental or intentional changes in the message provides integrity. The following figure illustrates the working of a MAC:

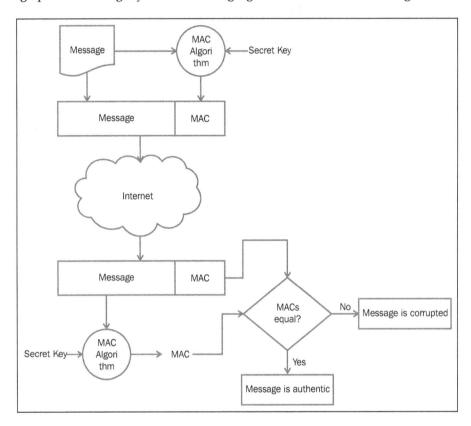

A MAC can be generated using different methods: by using a one time pad or one time secret key, by using a hash function, and by using a stream cipher or by using a block cipher and output the final block as a checksum. An example of the last method is DES with the CBC mode.

A hash function is used to create a checksum called **Hashed MAC (HMAC)**. This hash is then encrypted with a symmetric key and appended to the message. This is the most popular method of generating the MAC. Some examples of this kind of MAC are AES 128 with SHA1 and AES 256 with SHA1.

Android provides the capability to generate an HMAC by using the `javax.crypto.Mac` class. The following code snippet shows how to generate a digest with SHA-1:

```
String plainText = "This is my test string.";
String key = "This is my test key.";
Mac mac = Mac.getInstance("HmacSHA1");
SecretKeySpec secret = new SecretKeySpec(key.getBytes("UTF-8"),
    mac.getAlgorithm());
mac.init(secret);
byte[] digest = mac.doFinal(plainText.getBytes());
String stringDigest = new String(digest);
```

Summary

In this chapter, we discussed the tools that application developers can use to protect the privacy of their application and user data. We discussed about random number generation for seeding, and for use as initialization vectors for crypto algorithms. Hashing techniques, such as SHA-1 and MD5, were discussed, which developers can use to store passwords. They are also ideal to compress large data into a finite and defined length. Public key cryptography for the exchange of secret keys and symmetric key algorithms, such as AES to encrypt large amounts of data, were also discussed. We also discussed stream and block ciphers and block cipher modes. Most of the algorithms have test vectors published and available online. Developers can test their implementation against these test vectors. In the following chapters we will use these tools and techniques to protect data. Now let's move on to the next chapter to learn how to decide the best storage option for different types of data.

7
Securing Application Data

An application developer's credibility depends on how securely they handle their user's data. It is always prudent to not store huge amounts of user data on the device. It not only eats up memory but is a huge security risk as well. However, there are use cases where applications need to share data, cache application preferences, and store data on the device. This data could be private to the application or shared with other applications. An example of such data could be the user's preferred language or book category. This kind of data is kept by the application to enhance the user experience. It is useful within the application itself and is not shared with other applications. An example of shared data could be the wish list of books that the user keeps adding to the collection as the user browsers through the store. This data may or may not be shared with other applications.

Based on the privacy and kind of data, different storage mechanisms can be employed. An application can decide to use shared preferences, a Ccontent Provider, a file stored on internal or external memory, or even the developer's own servers to store data.

This chapter begins with the most important question of identifying information that an application should store and how to make a decision on storage location for data. As is always the case, a minimum amount of information should be collected and user consent should be obtained before collecting sensitive information. Next, we discuss storage mechanisms in Android, including shared preferences, device storage, external storage, and storing data on the backend servers. We will discuss protocols to secure data in transit. We will close the chapter with a discussion on installing an application on external memory.

Data storage decisions

A number of factors affect the data storage decision in the context of an application. Most of them are based on data security aspects that a developer should be aware of such as privacy, data retention, and implementation details of the system. These are discussed in the following sections.

Privacy

Applications today collect and use different kinds of information about their users. User preferences, location, health records, financial accounts, and assets are some of them. The collection of such information should be done with care and with user consent, as collecting private information can cause legal and moral issues and can be termed as invasion of privacy. Even when such information is collected, it should be stored properly encrypted and transmitted securely. Secure data storage and transmission is the focus of the later half of the chapter.

Privacy manifests itself in different forms. First, it is different in different cultures and countries. Each country has established rules and regulations about personally identifiable information, or PII. For example, the European Union has a **Data Protection Directive** for the processing and movement of personal data. More information can be found at this website maintained by the European Commission's Directorate General of Justice `http://ec.europa.eu/justice/data-protection/index_en.htm`. Cyber laws in India about this can be found at `http://deity.gov.in/content/cyber-laws`. The United States follows a sectoral approach to data protection. This is a combination of legislation, regulation, and self-regulation, rather than government alone.

Second, there are different laws for different use cases. For example, if an application is related to medical or health then the rules are different from an application that is tracking user location or making financial transactions. Some examples of specific laws in the United States are the Americans with Disability Act, Children's Online Privacy Act of 1998, and Electronic Communication Privacy Act of 1986. So, it is important to be aware of the rules and regulations pertaining to your use case and of the country in which you want to operate. When in doubt, use the services of companies that are experts in their field. For example, instead of trying to work out your own payment system, you might like to use a payment provider such as PayPal that has been doing payment processing for years and is compliant with rules and regulations such as PCI in this space.

Third, the transfer of private information from one country to another falls under rules and regulations as well. In most cases, the other country should have sufficient protection laws to satisfy the protection criterion of the other country.

The Universal Declaration of Human Rights, Article 12, states the privacy rule as follows:

> *"No one shall be subjected to arbitrary interference with his privacy, family, home or correspondence, nor to attacks upon his honour and reputation. Everyone has the right to the protection of the law against such interference or attacks."*

Some examples of PII include full name, e-mail address, mailing address, driving license, voter registration number, date of birth, mother's maiden name, birthplace, credit card numbers, criminal records, and national identification number. In some cases, age, gender, job position, and race may be considered as PII. Sometimes privacy may mean anonymity.

If your application is collecting PII, you will have to disclose it to your users and maybe take their consent. You can present them with terms and conditions for using the application or using certain features that might require your application to collect sensitive information about users.

Data retention

Data retention refers to the storage of data for the particular period of time. This data is used for tracing and identifying information such as people, devices, and location. For example, banking data is usually saved for seven years. Data retention should not be an issue in most use cases, unless it is an organization catering to a specific use case such as postal, banking, government, telecommunication, public health, and safety. In most cases, proper access rights have to be defined for access of this PII. Again, data retention rules are different for different countries and different use cases.

Implementation decisions

When dealing with data and deciding upon the most secure security mechanism, the first question is to identify where that data will be stored. Let's go back to our bookstore example. As we identified in *Chapter 3, Permissions*, the data elements of our example are:

- Name
- Credit card number
- Mailing address
- Last author searched
- Last language searched
- Last category searched

- Username
- Password
- Wish list of books

Further analysis of the preceding assets, based on their privacy needs, leads us to the identification of PII as name, credit card number, mailing address, and password. Please note that this classification is changed based on the country as well.

Next is the question of persistence. Do we want data to be available only during one instance of the application or multiple instances? Do we want the data to persist resets? In our example, we would like all the assets to be persisted. However, if the user preferences such as author, category, and language do not persist resets, we do not lose valuable information and the user can select them again.

The third important task is to identify which data is private to the application and which data is shared. Visibility of data will affect the storage option we choose.

The fourth question is that of size of data. Large files should preferably be stored on external storage. The following figure shows the memory options available in a typical Android phone device:

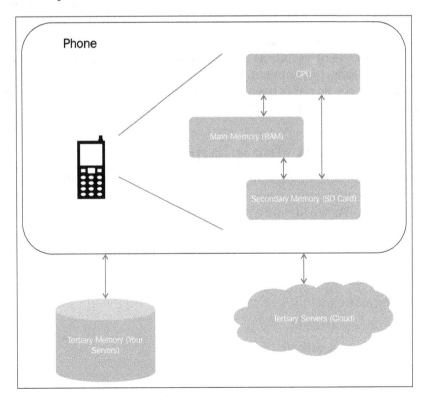

It is always advisable to use a storage mechanism provided by the framework instead of inventing a new one. In the sections that follow, I have discussed storage mechanisms provided by the Android framework for different storage needs.

User preferences

An application collects user preferences in two ways. In the first case, an application presents a settings screen to the user to choose preferences such as language, number of results to show per page, and so on. Such preferences are best stored using the `Preference` class. The other case is when user preferences are picked up as the user navigates through the application. For example, when searching for a book, the user selects books by a particular author. An application might want to save such a preference for the next time when the user logs in. Such user preferences are best stored using `SharedPreferences`. Under the hood, the `Preference` class also calls `SharedPreferences`. Please note that `SharedPreferences` only persist primitive data types.

Shared preferences

The `SharedPreferences` class is used to store primitive data types in a key-value pair. These primitive types include `int`, `long`, `Boolean`, `float`, `string set`, and `string`. Data stored in this `SharedPreferences` persists application sessions. A preference file is stored in the form of an XML file on the device in the `data` directory of the application. This file is thus sandboxed by the same Linux permissions as the application itself. The data in the preferences file persists even if the application is killed and is destroyed only when the application is uninstalled or specific values are removed using the methods of the `Preference` class.

Three operations for any kind of data storage are instantiating storage, storing data, and retrieving data.

Creating a preference file

The following code snippet instantiates `SharedPreferences`, using the default filename:

```
SharedPreferences preferences = PreferenceManager.getDefaultSharedPref
erences(context);
```

The filename in this case can be retrieved using the following code:

```
String preferencesName = this.getPreferenceManager().
getSharedPreferencesName();
```

You may specify the name of the preferences file as well. In the following example, the name of the preference file is `MyPref`:

```
public Static final String PREF_FILE = "MyPref";
SharedPreferences preferences = getSharedPreferences(PREF_FILE, MODE_
PRIVATE);
```

The preceding code snippet brings up an important discussion about preference file visibility and sharing. By default, all preference files are private to the application that created it. So their mode is `MODE_PRIVATE`. If a preference file needs to be shared among different applications, it can either be set to `MODE_WORLD_WRITABLE` or `MODE_WORLD_READABLE` to allow other applications to write and read the preference file respectively.

Writing preference

The next step is to store primitive data into a preference file. The following code snippet follows from the preceding code snippet and shows how to add data into the preference file. You will notice that you need the `SharedPreferences.Editor` class to store values. All values in the `Editor` class are batched and need to commit for values to persist. In the following example, `MyString` is the key for the string and its value is `Hello World!`.

```
SharedPreferences.Editor editor = preferences.edit();
editor.putString("MyString", "Hello World!");
editor.commit();
```

Reading preference

The next step is to read the key-value pairs that are in the preference file. The following code snippet follows from the preceding code snippet and shows how to read data from the preference file:

```
String myString = preferences.getString("MyString", "");
```

 SharedPreferences are accessible to all the components of the application. Other applications can write and read the preference file if set to MODE_WORLD_WRITABLE or MODE_WORLD_READABLE respectively.

To read the preference file of a different application, the first step is to get a pointer to the other application's context and then to read the value.

```
Context myContext = getApplicationContext().createPackageContext("com.
android.example", Context.MODE_WORLD_READABLE);
```

```
SharedPreferences preference =
myContext.getSharedPreferences("MyPref",Context.MODE_WORLD_READABLE);
String mMyString = preference.getString("MyString", "");
```

Preference Activity

With Honeycomb, Android extended the capabilities of the `Preference` class to collect settings from UI. These values are set as an XML file and the Activity inflates from it. Under the hood, the `Preference` class uses the `SharedPreferences` class to store key-value pairs. Such settings are private to the application and are only accessible by an `Activity` class.

To select a ringtone, the following code has to be set up in the `Preference.xml` file under the `res/xml` directory:

```
<RingtonePreference
  android:name="Ringtone Preference"
  android:summary="Select a Ringtone"
  android:title="Ringtones"
  android:key="ringtonePref" />
```

To inflate an Activity from this XML file, the following code is used in the `onCreate()` method:

```
public class Preferences extends PreferenceActivity {
  @Override
  protected void onCreate(Bundle savedInstanceState) {
    super.onCreate(savedInstanceState);
    addPreferencesFromResource(R.xml.preferences);
  .  .  .  .
  }
```

Remember to add this Activity in the manifest file.

File

An application can use Android's filesystem to store and retrieve data as well. The `java.io` package provides this functionality. This package provides classes to write and read different data types from a file. By default, files created by applications are private to the application and cannot be accessed by other applications. Files persist reboots and application crashes; they are only removed when the application is uninstalled.

Creating a file

The following code snippet shows how to create a file. As I have said before, by default, all files are private to the application.

```
FileOutputStream fOut = openFileOutput("MyFile.txt", MODE_WORLD_
READABLE);
```

The file `MyFile.txt` will be created in the /data/data/<application-path>/ files/ directory. The preceding file is created as MODE_WORLD_READABLE, that means that other applications can read this file. The other options are MODE_WORLD_READABLE, MODE_PRIVATE, and MODE_APPEND that let other applications write to the file, keep it private to the application, or append to it respectively. It is important to decide upon the appropriate visibility. As is always the rule with security, only give the least amount of visibility that is required.

Since MODE_WORLD_READABLE and MODE_WORLD_WRITABLE are very dangerous options to have, starting API level 17, these options have been deprecated. If files still need to be shared between applications associated with the same certificate, the android:sharedUserId option can be used. If these are different applications then file access can be handled with a wrapper class that interfaces with the file and provides read and write functionality. Access to this wrapper class can be protected using permissions.

Writing to a file

The next step is to write to a file. The following code snippet shows the use of the OutputStreamWriter class to write a string to a file. There are many options available in the java.io package to write different kinds of data to files. Please check the package to pick the correct option for your use case.

```
String myString = new String ("Hello World!");
FileOutputStream fOut = context.openFileOutput("MyFile.txt", MODE_
PRIVATE);
OutputStreamWriter osw = new OutputStreamWriter(fOut);
osw.write(myString);
osw.flush();
osw.close();
```

Reading from a file

Again as stated earlier, please check out the java.io package to find the best method to read data from a file. The following code snippet shows how to read the string from the file.

The following example reads one line at a time from the file:

```
FileInputStream fIn = context.openFileInput("MyFile.txt");
InputStreamReader isr = new InputStreamReader(fIn);
BufferedReader bReader = new BufferedReader(isr);
StringBuffer stringBuf = new StringBuffer();
String in;
while ((in = bReader.readLine()) != null) {
  stringBuf.append(in);
  stringBuf.append("\n");
}
bReader.close();
String myString = stringBuf.toString();
```

File operations on an external storage

A file can also be created on an external storage. If the API level is 8 or greater, Android provides a special function called `getExternalFilesDir()` to get the application directory on an external storage.

```
File file = new File (getExternalFilesDir(null), "MyFile.txt");
```

As you will notice in the preceding code snippet, the `getExternalFilesDir()` method takes a parameter. This parameter is used to identify the appropriate storage directory based on the media type. For example, to store a picture, `ENVIRONMENT.DIRECTORY_PICTURES` is used, and to store a music file, `ENVIRONMENT.DIRECTORY_MUSIC` is used. If such a directory does not exist, it will be created and then the file will be stored there. The value `null` is the root directory of the application.

```
File file = new File(
  getExternalFilesDir(ENVIRONMENT.DIRECTORY_PICTURES),
  "MyFile.jpg");
```

For an API level of less than 8, users can use `getExternalStorageDirectory()` to get the root of the external storage. The file can then be created in the `/Android/data/<application-path>/files/` directory.

To create a file on an external storage, the application should have the `WRITE_EXTERNAL_STORAGE` permission. The files created on external storage will be removed when the user uninstalls the application.

The external storage lacks the security mechanism of internal storage. It is best to assume that any data stored on the external storage is insecure and globally readable. And if external storage is not mounted, the file is not accessible and proper error handling mechanisms have to be employed for the application to fail gracefully.

In some cases, an external storage may actually be desired, especially if the files have no PII and are intended to be shared and be available on different devices. The media scanner will scan these directories when searching for relevant content. These directories are listed as follows. These follow the root directory of the application /data/data/<application-path>/.

- **Audio (music) files**: Music/
- **Podcast files**: Podcasts/
- **Video files (except for camcorder)**: Movie/
- **Ringtones**: Ringtones/
- **Pictures**: Pictures/
- **Miscellaneous downloads**: Downloads/
- **Notification sounds**: Notifications/
- **Alarms**: Alarms/

Cache

If an application needs to cache data, it is prudent to use a cache storage mechanism provided by the Android stack. Android stores cache files in the filesystem along with the application so that they are sandboxed with the application that created it. All cache files are created in the /data/data/<application-path>/cache/ directory. In case the system is running low on memory, these cache files are deleted first. Regular pruning of these files is necessary as they may grow big and eat up disk space.

The following code snippet first writes a string to the cache file and then reads the same string from the cache file. As you will notice, reading and writing is the same as any file input/output, only the location of the file is obtained using getCacheDir() to write a string.

```
//Write to the cache file
String myString = new String ("Hello World!");
File file = new File (getCacheDir(), "MyCacheFile");
FileOutputStream fOut = new FileOutputStream(file);
OutputStreamWriter osw = new OutputStreamWriter(fOut);
osw.write(myString);
osw.flush();
osw.close();

// Now read from the cache file
File file = new File (getCacheDir(), "MyCacheFile");

FileInputStream fIn = new FileInputStream (file);
InputStreamReader isr = new InputStreamReader(fIn);
```

```
BufferedReader bReader = new BufferedReader(isr);
StringBuffer stringBuf = new StringBuffer();
String in;
while ((in = bReader.readLine()) != null) {
  stringBuf.append(in);
  stringBuf.append("\n");
}
bReader.close();
String myString = stringBuf.toString();
```

Just as files can be created on an external storage, so can the cache files. The approach is different depending upon the API level. Starting with API level 8, Android provides a special function called `getExternalCacheDir()` to get the cache directory on an external storage.

```
File file = new File (getExternalCacheDir(), "MyCacheFile");
```

This directory is linked to the application and when the application in uninstalled this directory ceases to exist. If it is a multiuser environment each user has his/her own personal directory.

If the API level is less than 8, users can use `getExternalStorageDirectory()` to get the external storage, then create the file in the `/Android/data/<application-path>/cache/` directory.

To create a cache on an external storage, the application should have the `WRITE_EXTERNAL_STORAGE` permission.

Creating a cache on an external storage is not without security concerns. First, if the external storage has not mounted the cache file, it is not accessible and proper error handling mechanisms have to be employed for the application to fail gracefully. Secondly, an external storage is inherently insecure so anything stored on external memory should be assumed to be globally readable.

 The cache files should be pruned regularly and files that are not needed should be removed to preserve memory.

Database

A database is the best option to store structured data. Android provides support for SQLite using the `android.database.sqlite` package. This database is a part of the Android stack and the system administers the database. Using SQLite for mobile operating systems is a prudent choice as it is small and requires no setup or administration. And it is free!

Once created, the database files are sandboxed with the application and are stored in the /data/data/<application-path>/databases/ directory. This private database will be accessible to all the components of the application but not outside the application.

The following code snippet shows how to create a database that resides on the internal memory. The class will extend the SQLiteOpenHelper class and uses the SQL (Structured Query Language) CREATE_TABLE clause. The table stores a list of books that the user marks as a wish list. There are two columns in our table wishlist, a column ID that auto increments and the name of the book.

You will notice two methods here, onCreate() and onUpgrade(). OnCreate() will create a new database (if it does not exist) and a new database table. If the database already exists, the method onUpgrade() is called.

```
public class MySQLiteHelper extends SQLiteOpenHelper {
   public static final String TABLE_NAME = "wishlist";
   public static final String COLUMN_ID = "_id";
   public static final String COLUMN_BOOK = "book";
   private static final String DATABASE_NAME = "bookstore.db";
   private static final int DATABASE_VERSION = 1;
   @Override
   public void onCreate(SQLiteDatabase database) {
     database.execSQL("create table " + TABLE_NAME + "("
       + COLUMN_ID + " integer primary key autoincrement, "
       + COLUMN_BOOK + " text not null);");
   }

   @Override
   public void onUpgrade(SQLiteDatabase database) {
     database.execSQL("drop table if exists " + TABLE_NAME);
     onCreate(db);
   }
 . . . .
   }
```

Similarly, other database queries can be used to add a row, read a row, and delete a row. Any good book on SQL can help you with these queries.

It is also possible to create a database that exists in external memory. Creating a custom context class that accepts directory paths can accomplish this. You also need to have write access to the external storage. However, it is not advisable to do so if there is sensitive information in the tables.

As I noted earlier, the SQLite database is a private database and is sandboxed with the application. In case this data need to be shared with other applications then this is accomplished using a Content Provider that is addressed as a URI. We have already covered Content Providers in detail in *Chapter 2, Application Building Blocks*.

Account manager

In the context of storing sensitive data, storing passwords or authentication tokens is an important aspect. Consider applications such as Google Mail, Twitter, and Facebook, which lets users log in. Other applications work with an authentication token as used by identity protocols such as OAuth2.

Android provides the `android.accounts.AccountManager` class as a centralized repository for storing user credentials. An application can choose to use its own pluggable authenticator to handle account authentication. From storing the username to identity information to creating your custom account manager, Android's `AccountManager` is a powerful utility.

The `AccountManager` class functions are permissions protected so that your application will have to request for `android.permission.GET_ACCOUNTS` to access the list of accounts stored on it and `android.permission.ACCOUNT_MANAGER` for OAuth2.

Each account is in a namespace format. For example, a Google account uses `com.google` and a Twitter account uses `com.twitter.android.auth.login`. The `AccountManager` is accessed as follows:

```
AccountManager am = AccountManager.get(getApplicationContext());
```

The entire list of accounts can be retrieved using the following code:

```
Account[] accounts = am.getAccounts();
```

The `auth` token is acquired in the form of a `Bundle` and is retrieved using the named value of `KEY_AUTHTOKEN`.

```
String token = bundle.getString(AccountManager.KEY_AUTHTOKEN);
```

There are two important points to remember when using `AccountManager`. First, if your app is trying to authenticate using OAuth2, your application will be talking to the server so this may cause delays and the calls should be made asynchronously. Second, the credentials are stored on `AccountManager` in plain text. So on a rooted phone, these will be visible to any user using the `adb shell` commands. So, as is the case with storing information on a device, instead of storing passwords and other PII in clear, it should be stored in a cryptographically secure manner by wither hashing it or encrypting it. This will minimize the risk from a compromised device.

SSL/TLS

I was reading a very interesting research conducted by the students of the Leibniz University of Hannover and the Philipps University of Marburg, Germany, about MITM (man-in-the-middle) attacks on data in transit. The applications studied were using SSL (Secure Socket Layer) or TLS (Transport Layer Security) protocols to protect data over the network. Many of the applications were not using SSL/TLS properly, which resulted in vulnerability. Another interesting observation is that since the Android browser does not show the green padlock usually associated with the sites using SSL/TLS, users were not aware of the fact that they were using an insecure website. Check out the paper at `http://www2.dcsec.uni-hannover.de/files/android/p50-fahl.pdf`. I'm sure it will make an interesting read.

The preceding research brought to light the importance of implementing protocols correctly in the applications. This section introduces SSL/TLS and some notes to implement it correctly. Developed by Netscape, SSL is a protocol for secure communication over the Internet. The protocol follows a series of calls between the client and server where they negotiate on a key and cipher suite for data exchange.

Android provides the capability to integrate SSL/TLS using the `javax.net.ssl`, `org.apache.http.conn.ssl`, and `android.net` packages. The following figure illustrates the sequence in SSL:

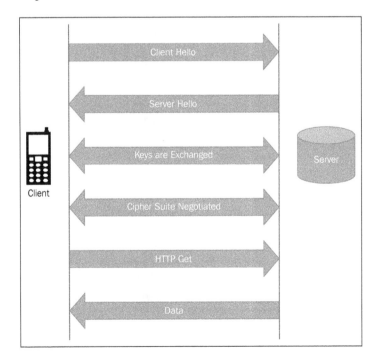

The first step is to set up a keystore and import the server certificate chain. Next is to link the keystore to `DefaultHttpClient` so it knows where to find the certificates for the server.

During the development phase, especially in an enterprise environment, we set up our SSL to trust all certificates by creating a custom `TrustManager` and to allow all the hostnames using `SSLSocketFactory.ALLOW_ALL_HOSTNAME_VERIFIER`. If such an application is released, it presents a serious security flaw. Please check for this before releasing your application. So remember to fix such issues before you release your application.

Installing an application on an external storage

As we have discussed in *Chapter 4*, *Defining the Application's Policy File*, starting with API level 8, applications can choose to be installed on the SD card. Once the APK is moved to external storage, the only memory taken up by the app is private data of applications stored on the internal memory. It is important to note that even for SD card resident APKs, the DEX (Dalvik Executable) files, private data directories, and native shared libraries remain on the internal storage.

Adding an optional attribute in the manifest file enables this feature. The **Application Info** screen for such an application either has a **Move to SD card** or **Move to Phone** button, depending on the current storage location of APK. The user then has an option to move the APK file accordingly. If the external device is un-mounted or the USB mode is set to **Mass Storage** (where the device is used as a disk drive), all the running Activities and Services hosted on that external device are immediately killed.

The following screenshot shows the option **Move to SD Card** in the setting of the application:

The `ApplicationInfo` object for each application now has a new flag called `FLAG_EXTERNAL_STORAGE`. The value of this flag is set to `true` for applications stored on the external devices. If such an application is uninstalled, the internal storage for that application is removed as well. If the external device becomes unavailable (for example, when the SD card is unmounted), the internal memory is not cleared. In this case, the user can clear this internal memory by uninstalling the application. The SD card does not need to be mounted for this action.

Two new broadcasts have been added as well.

- `ACTION_EXTERNAL_APPLICATIONS_UNAVAILABLE`: This intent is sent out when the SD card is unmounted. It contains a list of disabled applications (using the `EXTRA_CHANGED_PACKAGE_LIST` attribute) and a list of unavailable application UIDs (using the `EXTRA_CHANGED_UID_LIST` attribute).

- `ACTION_EXTERNAL_APPLICATIONS_AVAILABLE`: This intent is sent out when the SD card becomes available again. It contains a list of disabled applications (using the `EXTRA_CHANGED_PACKAGE_LIST` attribute) and a list of unavailable application UIDs (using the `EXTRA_CHANGED_UID_LIST` attribute).

When an application is moved from internal memory to an external location, `ACTION_EXTERNAL_APPLICATIONS_UNAVAILABLE` is fired. The assets and resources are then copied over to the new location. The application is then enabled and the `ACTION_EXTERNAL_APPLICATIONS_AVAILABLE` broadcast intent is fired again.

 Any kind of an external device is inherently insecure. For example, SD cards are susceptible to memory corruption due to power failure (dead battery in case of a phone) or improper removal of card (without properly unmounting). The SD card is also globally readable so applications can be read, written, copied, or deleted.

To securely store APKs on the external devices, Android applications are stored in an encrypted container (the ASEC file) so that other applications or programs cannot modify or corrupt them. The ASEC file is an encrypted filesystem whose key is randomly generated and stored by the device so that it can be decrypted only by the device that originally installed it. Thus, an application installed on an SD card works for only one device.

When mounting the SD card (using the Linux loopback mechanism), these containers are mounted in the same way as apps on the internal memory. The filesystem enforces permissions so that other applications cannot modify its contents and nobody but the system itself can modify anything through the ASEC file because other applications do not have the key for it. Also the SD card is mounted as `noexec` so nobody can put executable code there.

Multiple SD cards can be associated with one device so that SD cards can be easily swapped. As long as the SD card is mounted, there is no performance issue.

The Android developer website (`developer.android.com`) gives a list of use cases when installing an application on an SD card can make the application perform erratically, if the SD card is unmounted. Some of them, such as Services, are based on the sequence in which a Service will become available when a phone is booted up. These are listed as follows:

- **Services**: Running Services will be killed. The application can register for the `ACTION_EXTERNAL_APPLICATIONS_AVAILABLE` broadcast Intent, which will notify your application when applications installed on external storage have become available to the system. The Service can be restarted once the Intent is received.

- **Alarm services**: Alarms registered with `AlarmManager` will be canceled and must be manually re-registered when an external storage is remounted.

- **Input Method Engines (IME)**: An IME is a control that lets users input text. If your IME resides on an external storage, it will be replaced by the default IME. When an external storage is remounted, the user will have to open the system settings to enable the custom IME again.

- **Live wallpapers**: The default live wallpaper will replace running live wallpaper if the one that is set is stored on an external storage. When an external storage is remounted, the user will have to select their custom live wallpaper again.

- **App Widgets**: If your App Widget lives on the external storage, it will be removed from the home screen. In most cases, a system reset is required for the App Widget to appear again on the home screen.

- **Account managers**: If any accounts were created with `AccountManager`, they will disappear until the external storage is remounted.

- **Sync adapters**: `AbstractThreadedSyncAdapter` and all its sync functionality will not work. The external storage has to be remounted for sync to work again.

- **Device administrators**: This piece is critical as `DeviceAdminReceiver` and all its admin capabilities will be disabled, and this may not be fully functional even when the SD card is remounted.

- **Broadcast Receivers**: Any Broadcast Receiver that is listening for the `ACTION_BOOT_COMPLETE` broadcast will cease to work as the system delivers this broadcast before the external storage is mounted to the device. So any application installed on the external storage can never receive this broadcast.

Summary

This chapter covered the storage mechanisms available on Android. We started off with an understanding of terms such as privacy and data retention. We should always think about these issues before we collect personal identifiable information to avoid legal and moral issues. It is important to note that rules and regulations pertaining to privacy and data security are different based on country and use case. We looked at storing user preferences using shared preferences, storing, reading, and writing data on files, caches, and databases. We also covered some important considerations when using SSL/TLS and application installation on the external memory.

The next three chapters will cover very interesting topics such as device administration, security focused testing, and new and emerging use cases on Android. Keep reading!

8
Android in the Enterprise

With the omnipresence of mobile devices, increasing number of employees are bringing their devices to work, and are demanding that corporate data be accessible on their personal or enterprise mobile device. This is a great convenience but is also a challenge of increasing magnitude. The cost of corporate data loss due to a compromised or lost device is huge.

There are many challenges that the IT department faces with the proliferation of mobile devices in the workforce. The first challenge is the wide array of mobile devices with different form factors and capabilities. The second challenge is to get employees on board and accept enterprise control on specific applications and parts of their device. The third challenge is the ongoing support of device administration.

This chapter is focused on device administration of an Android device. If you are not developing applications for the enterprise, you may safely skip this chapter and move on to the next chapter about security focused testing of Android applications.

The chapter begins with the basics of device administration and unique challenges of an Android ecosystem. Second, we discuss the mechanics of setting up and implementing a device administration policy and a receiver for Android. We also discuss the security of the data stored on the device and in transit. We close the chapter with a proposal for the next steps for setting up device administration on Android and policy and compliance guidelines that device administrators should be aware of.

The basics

In the context of a device in an enterprise, three terms — BYOD, MDM, and MAM are used repeatedly. We will use them throughout the remainder of the chapter as well, so let us understand what each of them means.

The first term is **Bring Your Own Device (BYOD)**. This term refers to the recent trend of employees bringing their own mobile devices to work, and accessing corporate data and application on their personal devices. An example is accessing e-mail and office documents on personal mobile devices.

The second term that is used often is **Mobile Device Management (MDM)**. MDM refers to remote management of company owned or employee owned mobile devices that access enterprise applications and data. Features such as remotely wiping corporate data and asking a user to set a password are examples of MDM. These features enforce enterprise control over the system features.

The third term that is used frequently in this context is **Mobile Application Management (MAM)**. This term refers to the management of software and services on mobile devices that access the enterprise data. The examples of MAM include application upgrade, capturing crash logs and user statistics and sending it to the IT department. MAM is different from MDM as the latter focuses on device features, whereas MAM is focused on software and services installed on the device.

Understanding the Android ecosystem

Android is a challenging ecosystem with numerous customized releases. The following figure shows the usage of an Android version at the time of writing this book. As you will notice, at any given time there are different versions of an Android stack in use. Understanding the nuances and special needs of each version is a full time job in itself. You can always check the latest usage statistics available at `http://developer.android.com/about/dashboards/index.html`.

To add to the preceding problem, each manufacturer has a customized version of the Android stack with their chosen features and capabilities. On top of this stack, carriers add their customizations as well. This has created a highly fragmented market.

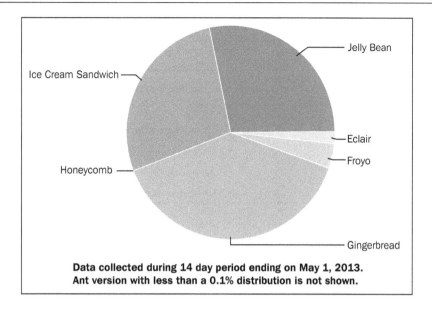

**Data collected during 14 day period ending on May 1, 2013.
Ant version with less than a 0.1% distribution is not shown.**

Device administration capabilities

Starting with Android 2.2, Android has added increasing capabilities to make
Android ready for enterprise. Each subsequent release has improved upon
existing capabilities and added more to it. The following table lists when a specific
enterprise capability was added in the Android stack. This section will focus on
some of these capabilities:

Android release	Enterprise features
Froyo (release 2.2)	• Password policy • Remote wipe • Remote locks
Gingerbread (release 2.3)	• SIP support
Honeycomb (release 3.0)	• Encryption and password policy for tablets • System encryption for tablets

Android release	Enterprise features
Ice cream sandwich (release 4.0)	• Extended system encryption, encryption and password policies to devices • Certificate management capabilities • VPN • Developer interfaces for SSL VPN • Facial recognition unlock • Network data usage monitoring • Offline e-mail searching

Device administration API

As illustrated in the preceding table, starting with Android 2.2, Android has continuously added support for device administration. The biggest step in this direction was the introduction of device administration API in Android 2.2 to support system-level control over devices that are required for enterprise.

The device administration API works in four steps:

1. The system administrator writes an application that enables policies to manage a device remotely.

2. The user downloads the application from Google Play or any other App Store. A user may also install the application using an e-mail.

3. Once downloaded, the user installs the application. At the time of installation, the user is presented with policies that will be enforced on the device. A user must agree to these policies to activate the application.

4. Once installed, the user must abide by these policies in order to access sensitive information. The user can uninstall the application, which will result in denied access to sensitive data.

The following figure illustrates the flow if the user has installed an admin application that enforces a password policy such that a password has to contain certain types of characters:

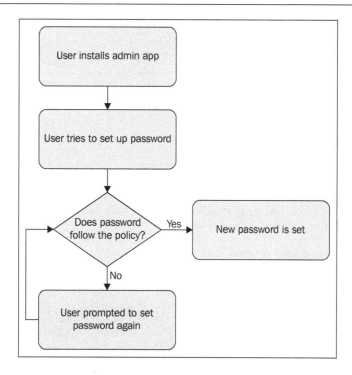

The device administration API is packaged as `android.app.admin`. This package has three classes: `DevicePolicyManager` for defining and implementing policies, `DeviceAdminInfo` that contains metadata for the device manager class, and `DeviceAdminReceiver` for implementing the receiver component.

Policies

Policies are an integral part of device administration. At the time of writing the book, the device administration API supported polices related to password, remote wipe, disabling camera, device encryption, and to lock the device. The examples of password policies include requiring a password to contain alphanumeric characters, password expiration and timeout, and maximum number of password trials. The list of current policies can be verified at `android.app.admin.DevicePolicyManager`.

A policy is defined in an XML file under the `res` folder. A sample policy file which imposes a limit on the password, can remotely reset the device to factory settings, disable a camera, encrypt storage, and lock the device. During install time these policies are shown to the user.

```
<device-admin xmlns:android="http://schemas.android.com/apk/res/
android">
    <uses-policies>
```

```
            <limit-password />
            <force-lock />
            <wipe-data />
            <expire-password />
            <encrypted-storage />
            <disable-camera />
        </uses-policies>
    </device-admin>
```

Additional policies are continuously added to new releases. You can check for the current version of the build and can enforce policies accordingly.

A device admin application contains `DevicePolicyManager` that manages policies for one or many device admin receivers.

```
DevicePolicyManager mDPMgr =
    (DevicePolicyManager)getSystemService
        (Context.DEVICE_POLICY_SERVICE);
```

Data can be remotely wiped from the phone with the following code. It is important to note that there are some fake device administration applications in the market as well. Be sure to download the correct admin application as suggested by your administrator. An unsafe or Trojan affected application can easily lead to data compromise:

```
DevicePolicyManager mDPMgr;
mDPMgr.wipeData(0);
```

To set the policy for encrypting filesystem, the following code snippet can be used:

```
DevicePolicyManager mDPMgr;
ComponentName mMyDeviceAdmin;
mDPMgr.setStorageEncryption(mMyDeviceAdmin, true);
```

DeviceAdminReceiver

Subclass the `DeviceAdminReceiver` class to create a device administration application. This class contains callbacks that are triggered when specific events happen. These Intents are sent by the system. Therefore the receiver should be able to handle the `ACTION_DEVICE_ADMIN_ENABLED` Intent.

`DeviceAdminReceiver` requires the `BIND_DEVICE_ADMIN` permission. `BIND_DEVICE_ADMIN` is a special permission that is accessible only by the system; applications cannot access it. This ensures that only the system interacts with the receiver.

The receiver also references the meta data policy file that we discussed in the previous section. The following code snippet shows a sample declaration.

```
<receiver android:name="MyDeviceAdminReceiver"
        android:label="@string/my_device_admin_receiver"
        android:description="@string/my_device_admin_desc"
        android:permission="android.permission.BIND_DEVICE_ADMIN">
    <meta-data android:name="android.app.my_device_admin"
                android:resource="@xml/my_device_admin" />
    <intent-filter>
      <action   android:name="android.app.action.DEVICE_ADMIN_ENABLED"
/>
    </intent-filter>
</receiver>
```

The following screenshots show the Exchange ActiveSync settings for corporate e-mails. This is just an example to illustrate the flow. In the first screenshot, the actual account details for exchange have to be filled in. These will be the corporate account details. Note the selection of the encrypted SSL connection:

In the next step, the user selects what functionality should be synced to the device. In our case, the user checks all the functionality provided by Exchange ActiveSync, namely **Mail**, **Contacts**, and **Calendar**. This is shown in the following screenshot:

In the third step, as shown in the following screenshot, a user has to confirm that they agree to the security policy that will be enforced on the device, if they decide to install the applications and be able to access the sensitive information. If the user declines to agree, the application will not be installed (in our case, **Mail**, **Contacts**, and **Calendar** will not be synced).

In the next step, the user reviews the policies that will be enforced by syncing the e-mail. These are the policies that are defined in the policy file, in the preceding example. In our example, as shown in the following screenshot, the device administrator can erase all the data on the employee device remotely in case the device is lost, an employee ceases to work for the enterprise, or for any other reason. The second policy is that the device administrator will set password rules. These password rules could be any of the following:

Protecting data on a device

A major requirement for MDM is to protect the enterprise data that lives on the device. An Android device usually has two forms of data storage: internal storage and external (removable) storage media. Starting with Honeycomb, an internal file system is mounted at `/mnt/sdcard` and an external storage is mounted at `/mnt/external#` (where # is the number of external devices). Earlier versions mounted the internal storage at `/mnt/sdcard` and a SD card at `/mnt/sdcard/external_sd`. Customized versions of the Android stack may or may not follow these guidelines.

Android addresses the problem of the enterprise data protection on a device by a full-disk encryption and by supporting the encryption algorithms.

Encryption

The ability to support full disk encryption was added in Android 3.0 to prevent unauthorized access to user data. The filesystem is encrypted using the `dm_crypt` kernel feature and works on a block device layer. The secret is derived from the user password and the encryption used is AES-128 with CBC and ESSIV: SHA-256. The master or the encryption key is encrypted by using Open SSL with AES-128.

For full disk encryption to work, a device needs to be protected with a password (the pattern password will not work). The device must be unlocked with the password before accessing the file system. The device administrator may set a policy for a limited number of password trials beyond which the device will reset to the factory settings.

A user must manually accept to encrypt the device. Note that when the device is being encrypted for the first time, the device should have sufficient power to complete the encryption process. If the device runs out of power, then it must be set to the factory settings and all user data will be lost.

[Only the file system on the device is encrypted. External memory, such as an SD card is not encrypted.]

As discussed in *Chapter 6, Your Tools – Crypto APIs*, the Android stack supports cryptographic algorithms, such as encryption and hashing. In cases where information has to be stored on the SD card, crypto functions supported by the stack can be used. A device administrator may enforce the policy that any data to be stored on the SD card has to be encrypted.

Even with full device encryption there are several issues that a user should be careful about. First is the case of shoulder surfing where people snoop the password over a person's shoulder in crowded places. One should be mindful of this issue. Second, although it is quite a hassle to enter the password to unlock the phone, it is for the security of the corporate data and instead of leaning towards choosing an easy password, it is advisable to choose a stronger password. The device policy may set a requirement for the same. Third, keep in mind that only data partition of the filesystem is encrypted. It is easy to store data elsewhere but any corporate data should be stored in data partition for security reasons.

Backup

Google provides backup services for Android devices. Data such as wallpaper, setting, dictionary, and browser settings are backed up. When the phone is set to factory reset, these setting are restored. Sensitive data, such as passwords, screen lock PIN, SMS, and call records are not backed up. Backup services can only be accessed using the `BackupManager` APIs. Backup has to be enabled manually by going to the **privacy** option under settings.

Google does not vouch for security of backup as different implementations of Android implement backup differently. This backup Service may not be available on all flavors of Android devices.

Secure connection

An Android supports VPN natively. An administrator may suggest a custom VPN and may require VPN to be turned on for all communication. This will come in handy specially when connecting over open hotspots. The latter facility is only available on Android 4.2. The following screenshot shows some VPN protocols supported on my phone:

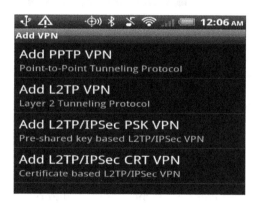

When connecting to a Wi-Fi, a user should choose a secured wireless connection. In this case, a user will be prompted to enter the password key. This is shown in the following screenshot:

Identity

Android supports a certificate store for storing certificates on a device and allows authorized apps to use it for identification in use cases such as e-mail, Wi-Fi, and **VPN (Virtual Private Network)**. Android supports DER-encoded X.509 certificates. It also supports X.509 certificates stored as PKCS#12 keystore files.

Android supports Bouncy Castle and comes pre-installed with certificates. They are available in the `cacerts.bks` keystore.

Users can also install certificates from their device memory. A new certificate can be installed on the device by navigating to the **Install from SD card** option in settings under the **Location & security** settings. Users should be mindful of what certificates they install as installing non-legitimate certificates may compromise the security of the device.

To remove a certificate, a user can go to **Personal | Security | Credential storage | Trusted credentials** and disable or remove the certificate.

Next steps

Now that we have an understanding of the Android capabilities to support BYOD, this section discusses how we can use this knowledge to roll out the Android support in an enterprise.

Device specific decisions

To use an Android device to its full capabilities, a device must be associated with a Google account. This enables the user to access Google Play, location services, and a host of other applications, such as Gmail, Drive, Calendar, and YouTube. The important question for device administrators is if they want employees to use their personal Google account or a separate enterprise account.

Another important question is the enablement of location services, which might be a privacy risk for some high value employees who might not like to be tracked. On the other hand, enabling location Service can help determine the location of the device if stolen.

Third important question is that of backup and storage. Again, as is the case with the location Service, backup and storage are important capabilities but might pose privacy concerns. The device administrator might enforce encrypted storage or designate a private enterprise cloud. But this soon adds up to maintenance costs. To enable backup, a user must explicitly go to **setting | privacy** and select **Back up my data** as shown in the following screenshot:

Here, it is important to address the question of the rooted Android devices. To root an Android device does not take much time and the instructions are readily available. Rooting a device is also legal in Australia, Europe, and the United States. A rooted device fails security standards for enterprise use. Thus, the detection of a rooted device is an important consideration for a device administrator. It is not easy to detect rooted devices, as there are many ways to root a device.

Then, there is the question of App Store from where employees download enterprise applications. Besides Google Play, Android application can be downloaded from other App Stores such as the Amazon App Store and GetJar. Last time I checked there were over 128 App Stores and an application can be hosted on any of them. Applications can also be downloaded from websites or via e-mail or through side loading. A device administrator may choose to set up an enterprise App Store to address this situation. This ensures that only legitimate applications exist here. To enable an application download from the external locations besides Google Play, a user must explicitly select the **Unknown Sources** option as shown in the following screenshot:

The basic idea of a device administration is that the device should be usable, intuitive, and should preserve the native experience without compromising security. It is a hard balance and is a sure challenge.

Keeping a tab on Android's evolving and vibrant ecosystem requires an Android expert and enthusiast who is passionate about Android and keeps abreast of the upcoming changes in the Android ecosystem. To keep your knowledge current, keen understanding of how users are interacting with their devices and upcoming innovation in this space is required. This Android expert should be an authority and the main point of contact for an enterprise application deployment on the Android devices.

Knowing your community

The second important step in this roll-out is to understand the preferences, requirements, and needs of your employee base. This step is important to make an informed decision about what applications and services are required by your employees and the kind of access control and policy that needs to be created. Gathering information about their device preference (whether they like phones or tablets), the applications they are comfortable with, the amount of access they require on their devices is important. Another factor is the geographical diversity. There is no one-size-fits-all solution. Different locations have different preferred devices, their own preferred applications, and their own level of interaction with the corporate data on a device.

Defining boundaries

Defining a clear boundary for which devices are accepted and which are not will help with dealing the fragmentation issue on Android. These boundaries should be based on capabilities and not on versions or releases, since manufacturers and carriers port the same version differently on different devices.

Another boundary to define is that of trust. A company's **Information Technology (IT)** department should allow increased access based on an increasing level of device capability. For example, if a device does not support full disk encryption, they can only read data and cannot store it on the device. Owing to Android's open app ecosystem, regular monitoring of application that users install on their device is important as well.

The third boundary is that of application that a user can install on their devices. Android applications can be installed from different sources and these sources do not curate the applications for security purposes as closely as the Apple App Store. Defining a boundary for which applications are allowed and which ones are not will go a long way in keeping the device secure.

Android compatibility program

Openness is the goal of an Android ecosystem. However, for a consistent user experience on different devices, the OEMs must participate in an Android compatibility program. This program provides tools and guidelines to the OEMs so they can correctly label their devices and to ensure applications run as expected on a device. This is an interesting program for IT personnel as they can define their boundaries based on the compatibility level.

The compatibility program provides three key components:

- **Compatibility Definition Document** (CDC): This is the policy document for compatibility. It defines the requirements of a compatible stack. For example, it lists a set of Intents that are considered as core to the Android stack and should always be supported.

- **Compatibility Test Suite** (CTS): CTS is a free test suite that runs on a desktop and can be used to automatically run compatibility tests on an emulator or device. At the time of writing this book, CTS included unit tests, functional tests, and reference tests with robustness, and performance tests planned for future. Some examples include checking for hardware features, such as Wi-Fi and Bluetooth.

- **Compatibility Test Suite Verifier** (CTS Verifier): CTS is a free test suite that runs on a desktop and needs manual input to run compatibility tests on an emulator or device. It is a supplement of CTS.

Based on the preceding criteria, three types of Android devices exist in the market. These are shown in the following table with key characteristics of each compatibility type:

Google lead devices	Google experience devices	Other (open) devices
<ul><li>Pure Android 100 percent Google</li><li>No OEM or carrier customization</li><li>Examples: Samsung Galaxy Nexus, Motorola Xoom, HTC Nexus Ones</li></ul>	<ul><li>CTS compliant</li><li>OEM and carrier customization</li><li>Should meet Google upgrade promise</li><li>Examples: Samsung Galaxy S11, HTC Rezound</li></ul>	<ul><li>Not CTS compliant</li><li>Highly customized by OEM and carrier</li><li>Examples: Kindle Fire, Motorola ET1 tablet</li></ul>

You may decide to only support lead and experience devices that pretty much provide consistent features and somewhat a customized experience.

Rolling out support

Plan out a phased approach to roll out support for Android devices. The IT department can start with a pilot roll-out and then expand it slowly. This helps in two ways. First, IT can determine if their support infrastructure can scale with the increased number of users. Second, they can adjust their support based on the usage statistics gathered. Any errors and missing requirements can be fixed as the support expands to more employees.

Educating employees through training, wiki, posters, and alerts during this roll-out will help employees understand what is happening. This also helps them understand the rationale behind why some devices are allowed and others are not, what to expect, and safe practices to access corporate data on devices.

Policy and compliance

While reviewing all the preceding steps, do not sideline the emerging standards and compliance in this area. Also, keep abreast of research in the field of BYOD, MDM, and MAM, and novel approaches being adopted by different firms.

FINRA

The **Financial Industry Regulatory Authority (FINRA)** is the largest independent regulator for all security firms doing business in the United States. FINRA's mission is to protect America's investors, by making sure the security industry operates fairly and honestly. They have issued guidelines regarding the supervision of electronic communication from mobile devices at its member firms. These need to be considered in conjunction with a firm's own analysis. Check out FINRA's website for more information: `www.finra.org`. There are three notices that FINRA has issued to expand upon the increasing proliferation of personal mobile devices and social websites. In all cases, it suggests that proper training should be given to all employees for it, maintaining records, judicious about posting on social media websites, and continuous supervision.

FINRA issued their first regulatory notice 07-59 in December 2007 (`https://www.finra.org/web/groups/industry/@ip/@reg/@notice/documents/notices/p037553.pdf`). This notice provides the core guidelines on supervision of electronic communication via mobile devices. It suggests that corporate e-mail should always flow through corporate e-mail systems and should not be forwarded through personal accounts. These corporate e-mails should only flow through monitored networks. This will enable proper supervision of e-mails.

FINRA's second regulatory notice 10-06, issued in January 2010, focuses on the use of social media websites and blogs (`http://www.finra.org/web/groups/industry/@ip/@reg/@notice/documents/notices/p120779.pdf`). This notice suggests that employees should not use business accounts on social media networks. These sites should be screened continuously for employee representation, as misleading information can adversely affect investors.

The third regulatory notice 11-39 issued in August 2011 expands on the guidelines on personal devices and social media sites (`http://www.finra.org/web/groups/industry/@ip/@reg/@notice/documents/notices/p124186.pdf`). This notice states that employees can use personal devices for communication as long as this information is retrievable and separate from personal communication. Ongoing supervision of devices is critical, as is ongoing training.

Android Update Alliance

It is not always easy to comply with standards. At Google I/O in May 2011, Google, along with many other device manufacturers promised to update devices within 18 months of any new release of Android. This alliance was called the Android Update Alliance. The idea was noble and very much appreciated, but it has been hard for the OEMs to keep up with it.

Summary

In this chapter, we focused on the management of company and employee owned devices that access the enterprise data. The issue for BYOD is that of trust, compliance, governance, and privacy, as more and more employees demand access of corporate data on their mobile devices. It is a delicate balance between user experience and security. We started off with the challenging Android ecosystem, followed by implementation details of device administration and a discussion of other enterprise capabilities provided by the Android stack. We closed the chapter on the next steps that should be considered to start supporting Android in the enterprise space with a discussion around compliance and policy.

Now, it's time to move on to the next chapter that discusses testing the Android applications from a security perspective. Happy reading!

9
Testing for Security

This is undoubtedly the most important chapter of the book. As developers, we all try our best to write beautiful, usable, and secure code. We have all experienced the thrill of a great idea and the rush to see it working. We also have crazy schedules and deadlines. So bugs happen and testing for bugs is a natural part of any coding lifecycle.

Most of the test cases today focus on usability, functionality, and stress testing. In most cases, test engineers are at a loss when it comes to testing for security. When compliance and security is overlooked, sometimes the application needs to be redesigned or implemented again. Take the case of creating a message digest for integrity purposes. A developer may decide to go with SHA-1 that creates a digest of 160 bits. On the server side, the database is designed to accommodate 160 bit data. A non-ethical hacker breaks into the application. When the security review is performed, it is decided that SHA-1 was not strong enough for the use case and needs to be updated to SHA-256. Since the database was designed to accommodate only 160 bits, it becomes a challenge to make a quick fix on the client side, as the entire design has to be changed. Things become serious now. This is such a waste of time specially considering the fact that a mobile ecosystem is fast paced and transient.

This chapter aims to introduce the concept of security-focused testing. The chapter begins with an overview of testing. If you are already familiar with testing, you may easily skip this section. The next section will talk about security testing and ways that you can test your applications for security, namely security review, manual testing, and automated testing with tools. The following sections discuss some sample security test cases that can act as a baseline for writing tests. The chapter closes with a discussion of tools and resources that developers and test engineers can use for development of test cases and for testing for security.

Testing overview

With a wide array of devices of varying capabilities, form factors, and versions, Android is one of the most challenging operating systems to test. Getting the basic functionality and user experience itself is a challenge. The following figure illustrates tests that are usually performed in the context of an Android application development. As Bruce Schneier, a great cryptographer of our times, aptly states, "Security is not a product but a process", so you will notice that I have added security testing to the entire lifecycle of application testing.

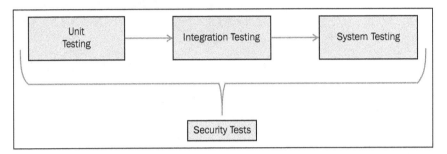

Let us spend a little time on what each category, unit testing, integration testing, and system testing means from an Android perspective.

- **Unit testing**: In most cases, developers who code the module develop unit tests. Developers should write and unit test their modules before handing off their code to test engineers. The Android SDK comes bundled with instrumentation APIs for unit test. This framework is implemented on JUnit, a popular framework for Java unit tests. Unit tests can be easily automated. These tests cover boundary tests, input validation tests, and connections with the backend.

- **Integration testing**: Once unit tests are done and different components are being integrated, integration tests are performed to make sure that different components work together. These are tests that are performed when components are bundled together. Let us consider two teams working separately, one on the login module and the other on the search result page. Once the development of modules is done and they are integrated with each other, tests should be performed which check the two modules together. These days most development environments use continuous integration that perform sanity tests that the two modules compile together.

- **System testing**: These are tests that test the whole application and how the application interacts with the Android platform. Some examples of system testing will include testing how search functions on different platforms and how differences in Android-based devices affect the display of search results.

Security tests should be performed at each stage of testing. For example, at the unit test level, developers should test for inconsistent and incorrect input values, buffer overflows, and access level of users.

At the integration level, engineers can test for secure data transit between two modules and behavior on passing erroneous data.

In the system testing state, engineers can test how their application looks and behaves on different Android platforms. In the case of Android, this is an especially important testing phase due to variations in the Android device and stack capabilities of different vendors and carriers.

Any of the previously mentioned test suites in the flow usually contain a mix of different kinds of tests. These are illustrated in the following figure. Notice that I have again added security tests in the mix.

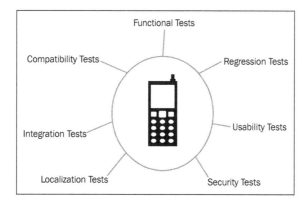

- **Functional tests**: These tests check if an application performs with expected behavior. For example, a functional test for the login function will test the case when a user enters a username and password and presses *Enter*; the user is either logged into the system if the credentials are valid or is shown an error. You may like to validate that correct error messages are generated with different error cases.

- **Localization tests**: Most of the applications today are global and are available in different countries. To support different locales, an application has to be localized and internationalized. Localized refers to language translation and internationalization refers to adapting the application according to the norms of a particular locale. Consider a case of adapting a view that accepts addresses in Japanese (that is, you want to support Japan as a country). Localization will translate the word Address Line 1, Address Line 2, City, State, Zip, and Country into their Japanese equivalent. However, in Japan the addresses system is different from the Roman system and the view that accepts address will have to be redesigned and some labels might have to be shuffled around.

Android has a very user friendly framework to store strings and localized views that developers should take advantage of. It is better to consult with localization experts when opening up an application in new markets.

- **Usability tests**: Also called UI tests, these tests focus on the look and feel of the user interface and make sure that it is easy for a user to enter input, read information on screen, and change the esthetics and general flow of an application. Usability is very important in a screen space constrained device and devices of different screen sizes.

- **Hardware compatibility tests**: This suite of tests will be targeted at testing hardware features used in the application on different devices. For example, if an application uses the device camera, tests should be performed to check if the code works properly on different device cameras with different focus capabilities.

- **Regression tests**: These are usually automated tests that are run after every change in the application to ensure that the application still works as expected. For example, in the bookstore application, you may identify key pieces of functionality as login, logout, searching for books, and adding a book to a wish list. Whenever a new feature is added or an existing feature is updated, these sanity tests are performed to make sure that nothing broke.

Security tests are discussed in detail in the following sections.

As you may have guessed already, most of these test cases work in tandem with each other. For example, to test the address page for a new country, both localization and UI tests have to go hand in hand.

Security testing basics

This chapter is an overview of security testing. We discuss the pillars of security around which security tests can be developed. The second section discusses different kinds of security testing.

Security tenets

Any kind of application security testing should follow the six tenets of security, namely authentication, authorization, availability, confidentiality, integrity, and non-repudiation. We have covered most of these concepts and how to implement them in *Chapter 6, Your Tools – Crypto APIs*.

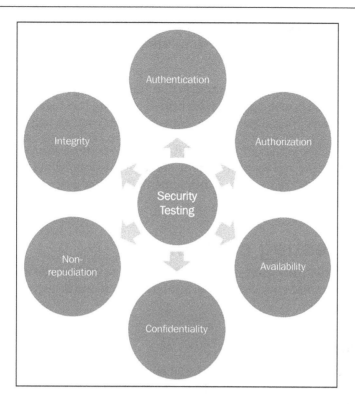

Authentication is the measure of identifying the user. You can use authentication APIs from companies such as Facebook, Twitter, LinkedIn, and PayPal. The main protocols used are OAuth and OpenIDConnect. These techniques offload the task of authentication from the application. It is a win-win condition for both application developers and users. Application developers do not have to implement their own schemes and can use built-in mechanisms for authentication. Users do not need to share personal information with applications they do not trust. It is useful for companies that develop such schemes as it drives traffic to their websites. Most of these techniques are based on providing a user with an authentication token.

Chapter 10, *Looking into the Future*, discusses some new advances in authentication.

Authorization refers to access control — determining whether the user has appropriate rights to access a resource. In case of Android, this can be achieved by protecting application components with permissions and checking caller identity whenever possible.

Availability means that data should be available to authorized users when needed. Using broadcasts and Intents with data can ensure this security measure.

Confidentiality refers to keeping data secure and only revealing it to intended parties. Encrypting data, using proper permissions, and conforming to the Android sandbox can help with this measure of security.

Integrity means that the data is not modified in transit or at rest. In case data is tampered with, this tampering can be identified. Adding message digests, digital signatures, and encrypting data, both in transit and at rest, can help with integrity.

Non-repudiation can be achieved using digital signatures, timestamps, and certificates, and ascertains that the sender cannot deny sending the data. DRM, discussed in *Chapter 6*, *Your Tools – Crypto APIs*, is implemented so that users cannot deny the reception of content.

Security testing categories

Keeping the security principles just discussed in mind, security testing can be categorized into three buckets: application review, manual testing, and automated testing.

Application review

The first step in security testing is the application review process. This process focuses on understanding the application and identifying hardware, different technologies, and capabilities that an application uses. Once these characteristics are identified, the reviewer tries to access the security holes in these capabilities. The review process identifies the issues that are pronounced in the manifest file, use of broken or weak cryptography, insecure use of protocols, and security issues in technologies and hardware that may have escaped during development. It covers compliance and standards and if they have been appropriately adhered to.

Some examples of security issues that can be identified in the manifest file include superfluous permissions in the manifest file that the application does not need but were added for debugging purposes, not protecting components using permissions, forgetting to turn off the debug mode, and log statements.

Compliance is based on the use case. Different standards are applied based on the use case around which the application is written. For example, a payment and commerce use case application may be looking at PCC-DSS. A geo-location based application will have to be aware of privacy issues.

These days there are security-auditing firms that specialize in the mobile application review process. You should use them if in doubt.

Manual testing

Manual security testing, as the name implies, is done manually during development or by test engineers. Engineers watch the behavior of an application under different scenarios by feeding in different inputs. Examples include looking at the logs to verify no sensitive information is being leaked out, going to the previous Activity several times to see how the application performs, trying to break the authentication scheme of the application, and checking if a user has appropriate access. Scenarios that cannot be also belong to this category. There are some companies such as uTest (`www.utest.com`) that employ manual testers who can perform manual testing for your applications.

Dynamic testing

Also called automated testing, these tests are ideally performed by writing test scripts. Tests such as input validation, stress testing, fuzzing, and boundary tests can be easily automated. Most of these tests can easily be part of the standard development/test cycle and can act as sanity tests when adding new features. You may decide to use services of security firms such as Device Anywhere (`www.deviceanywhere.com`) that specialize in this area.

Sample test case scenarios

In this section, I have tried to enumerate some sample test cases that are interesting from the security perspective. They are in no particular order and you may use them as a reference when identifying test cases for your particular use case.

Testing on the server

The mobile ecosystem is very interesting; it is young and still evolving. The application may want to send a piece of data to the server but what is received on the server may be quite different. This could be because of issues in the communication channel where a hacker snoops in and changes the content as it moves or it could be a bad client. No matter what the reason is, it is not enough to test the application only, server testing is critical for the security of your application. These tests focus on whether what was intended is received on the server side, if PII is stored in clear on the server, if business logic is residing on the client then is it working properly. We discussed this in *Chapter 6, Your Tools – Crypto APIs*, as well.

This field of testing is mature and plenty of examples and tools are available for testing on the server. Checking for open ports and firewall can be easily done using port scanning tools such as Nmap.

Testing the network

The infrastructure layer is the backbone of mobile devices and makes mobility ubiquitous. It also brings in new challenges and test cases. Devices communicate with the server using different protocols and each brings with it unique bag of security holes. GSM can be easily broken; Wi-Fi is inherently insecure, especially if you are connecting to a rogue hotspot. **Long Term Evolution (LTE)** is a new standard for high speed wireless data communication and is IP-based but has not been thoroughly tested, and the proximity technologies such as NFC, Bluetooth, and RFID bring a whole different testing paradigm. It is thus important to test the technology that your application is using and build test cases around it.

Securing data in transit

It is good if your application uses **Transport Layer Security (TLS)**, but make sure it is implemented correctly, so test it out. Testing that all communication between the client and the server is encrypted and no PII or secret key is being transmitted in clear. Remember serialization is not encryption and obfuscation is not encryption. Make sure that the server is checking for certificate validation and certificate expiry. Check if the encryption algorithms and protocol being used are current and secure enough for your use case.

Secure storage

It is always a good idea to not store sensitive data such as private keys, usernames, passwords, and other PII on the client. Ideally, this information should be stored on the server. Storing the key with the data it encrypts defeats the purpose of security. If keys have to be stored on the client, first they should not be stored in clear. Second, they should not be stored in files, cache files, or shared preferences. Keys should be stored in the `keystore`, passwords in `AccountManager`, and all sensitive information should be stored in an encrypted fashion. In most cases, a hash value can be stored instead of a password.

Validating before acting

Validate the input, the data, and the caller being passed between different components of the application and also from other applications. Any Activity can parcel any type of data in the Intent and it is the responsibility of the receiving component to test and verify before acting on it. We discussed this at great length in *Chapter 2, Application Building Blocks*. The tests in this case will include passing invalid and erroneous data to a component and observing how it behaves.

In some cases, you might be able to check the caller identity before acting on a request from them. Use it! Especially before launching sensitive actions, check the caller identity and data that you will be working on.

The principle of least privilege

Tests in this category include testing permissions of different application components and making sure that they have the least privileges possible to function properly. This includes checking files, cache files, and `SharedPrferences` for visibility and accessibility permissions. Check if they really need a `MODE_WORLD_READABLE` or `MODE_WORLD_WRITABLE` permission.

Check the permissions that your application requests. For example, if you do not need fine-grained location access, only ask for coarse-grained permission, and if you only need to read SMS, do not ask for both read and write SMS permissions. As consumers get more aware of security issues in the mobility space, they might be suspicious of your application if it requests permissions that do not make sense. It makes no sense for a book browsing application to have access to the user contact list and device camera.

Managing liability

Be aware of rules and regulations in your domain. Getting into liability lawsuits is a messy affair and it is best to stay away from them. Also, if it makes sense to use third party tools and services that specialize in these matters then by all means use them. If your application collects user data, make sure that you have taken proper consent from the user and everything that you are collecting is listed out. As an example, the California Online Privacy Protection Act states that if an application collects information in California then it should be disclosed.

Let's take another example use case of an application that processes payments. Instead of trying to do your own, use existing payment solutions such as PayPal. Payments deal with users' money and there are directives such as PCI-DSS that governs how such capabilities can be used.

Similarly, instead of designing and developing your home-grown security algorithms and protocols, use time tested and industry tested security suites and libraries.

Be aware of how the application will be used in the countries that you support. Different countries have different rules and regulations. The definition of PII is different as well.

Cleaning up

First, do not log sensitive information. Before releasing the application in the wild, make sure to turn debugging off. Clean up all sensitive information from files, cookies, and caches; that is, zeroize memory.

Usability versus security

Balancing usability and security is a tricky and delicate art. An application may persist username, password, and session tokens for convenience but it also makes security weak. If you have some functionality in your application that remembers user identity, weigh convenience and security. You may decide to limit session lengths and limit how long you want to keep cookies and tokens alive.

Authentication scheme

The question here is whether you want to authenticate the device or the user? Devices do get lost or stolen. Identifying a user based on device characteristics such as IMEI, IMSI, or UDID may not be such a great authentication scheme. These persist remote wipes and resets. You may like to evaluate biometrics-based authentication mechanisms or two factor authentication schemes to authenticate the user.

Thinking like a hacker

Think like a hacker and test how a hacker will try to hack your application. Use the available tools and exploits already listed on the Internet. Testing the application using tools used by hackers can reveal what hackers will see and what information they will get when they try to crack your application. There are tools such as Fiddler (www.fiddler2.com) that you may use to monitor the network traffic through your application. It is important to remember that obfuscating the code is not security.

Integrating with caution

Whether you are integrating with hardware (both internally and externally) or third party applications, do it with care.

If the application is using some hardware components such as camera, Bluetooth, NFC chip, accelerometer, microphones, or GPS then it is important to test them for security as well. A flaw in any piece of hardware can affect the overall security of the application.

Similarly, a bug in a third party library can result in the application being compromised. When integrating with such an external library, ask for their test results and look them up online and ask for recommendations.

Security testing the resources

This section focuses on tools, techniques, and some other resources that can creatively be used to test applications for security.

OWASP

OWASP (Open Web Application Security Project) is an organization committed to mobile security. They provide tools and research in the field of mobile security. Check out their website at `https://www.owasp.org`. It is a good place to look for your security-related questions, contribute to the open source and innovate and participate in the mobile security discussion. OWASP compiles a list of top 10 security vulnerabilities each year and the community is challenged to work on it.

Android utilities

Android provides a whole array of utilities that can be creatively used to test applications. Besides testing, these tools can also help developers debug their applications.

Android Debug Bridge

Android Debug Bridge (ADB) can be used for logs, memory checking, and many other purposes. Check out the full listing of functions provided by ADK on the developer website. The following screenshots show some examples of ADB in action:

```
                        platform-tools — adb — 80×24
LM-SJN-00713218:platform-tools prarai$ ./adb logcat
I/AccountTypeManager(  313): Loaded meta-data for 1 account types, 0 accounts in
 275ms(wall) 25ms(cpu)
D/dalvikvm(  257): GC_CONCURRENT freed 216K, 4% free 9202K/9543K, paused 5ms+21m
s
D/Launcher.Model(  168): Reload apps on config change. curr_mcc:310 prevmcc:0
E/WindowManager(   77): Window Session Crash
E/WindowManager(   77): java.lang.IllegalArgumentException: Requested window and
roid.os.BinderProxy@415a3d80 does not exist
E/WindowManager(   77):         at com.android.server.wm.WindowManagerService.wi
ndowForClientLocked(WindowManagerService.java:7163)
E/WindowManager(   77):         at com.android.server.wm.Session.setWallpaperPos
ition(Session.java:360)
E/WindowManager(   77):         at android.view.IWindowSession$Stub.onTransact(I
WindowSession.java:419)
E/WindowManager(   77):         at com.android.server.wm.Session.onTransact(Sess
ion.java:111)
E/WindowManager(   77):         at android.os.Binder.execTransact(Binder.java:33
8)
E/WindowManager(   77):         at dalvik.system.NativeStart.run(Native Method)
D/dalvikvm(  168): GC_CONCURRENT freed 337K, 4% free 9972K/10375K, paused 3ms+6m
s
I//system/bin/fsck_msdos(   31): Attempting to allocate 119 KB for FAT
I/ActivityThread(  348): Pub com.android.calendar: com.android.providers.calenda
```

The previous screenshot shows the sample log using the `adb logcat` command.

Setting up the device

To set up advanced settings for monitoring web applications, you may turn on advanced features that can help you get more data during penetration testing.

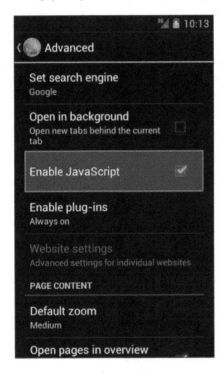

The previous screenshot shows how you may enable the JavaScript and plugin to check for information leakage.

SQlite3

Using the SQLite3 utility, a user may explore the databases it creates and some other databases that come bundled with the platform.

The SQLite utility lets the user query databases and check values in the database. Such database interrogation and examination can point out issues such as storing PII in the clear.

Dalvik Debug Monitor Service

Dalvik Debug Monitor Service (DDMS) is another important tool provided by the Android framework. DDMS provides capabilities such as port-forwarding, screen capture, thread and heap information, the `logcat` process, and radio state information, incoming call and SMS spoofing, and location data spoofing. The following screenshot shows the window for DDMS. Check out the details on capabilities on the Android developer website.

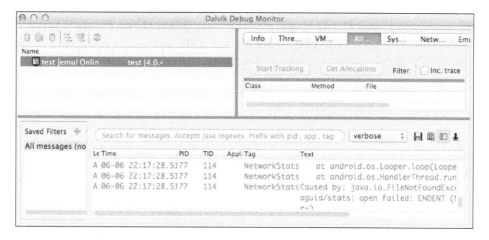

There are some other third party tools as well such as the Intent Sniffer and the Manifest Explorer, both developed by iSecPartners (https://www.isecpartners.com). Other Linux tools such as `strace` and `procrank` can also be used as well. You can use BusyBox, as discussed in the following section, for this purpose.

BusyBox

BusyBox is called the Swiss Army Knife of embedded Linux and provides several Unix tools such as `vi`, `whoami`, `watchdog`, and so on. These tools can be used for testing without rooting the phone. Installing the BusyBox on Android is pretty simple. Just download it from www.busybox.net.

```
LM-SJN-00713218:platform-tools prarai$ ./adb shell
# mkdir /data/busybox
# exit
LM-SJN-00713218:platform-tools prarai$ ./adb push ~/Downloads/busybox /data/busybox
2742 KB/s (1096224 bytes in 0.390s)
LM-SJN-00713218:platform-tools prarai$ ./adb shell
# cd /data/busybox
# ls
busybox
# chmod 777 busybox
# ./busybox --install ./
# ./vi hello.txt
```

As shown in the preceding screenshot, `busybox` can be easily pushed and installed. Once installed, the Linux commands can be easily executed.

Decompile APK

It is relatively easy to decompile an APK and read its contents. Doing this exercise will help you understand how a hacker will approach your APK.

An APK file is nothing but a ZIP file and renaming the APK file as a ZIP file will let you open it with any ZIP file explorer. These are available under the /data/app directory. You can pull it onto your machine using the `adb pull` command. In there, you can see the manifest file, resources, assets, and others.

```
LM-SJN-00713218:platform-tools prarai$ ./adb pull /data/app/ApiDemos.apk
3120 KB/s (2720164 bytes in 0.851s)
LM-SJN-00713218:platform-tools prarai$
```

Next, using the `dexdump` utility provided in Android, the classes under the /data/dalvik-cache directory can be dumped.

```
LM-SJN-00713218:platform-tools prarai$ ./adb shell
# cd /data/dalvik-cache
# ls
data@app@com.example.example1-1.apk@classes.dex
data@app@com.example.example2-2.apk@classes.dex
data@app@com.mycompany.example-2.apk@classes.dex
data@app@com.paypal.android.interactivedemo-1.apk@classes.dex
data@app@com.paypal.android.pizza-2.apk@classes.dex
data@app@com.paypal.android.simpledemo-2.apk@classes.dex
data@app@com.paypal.example.android.ppaccess-1.apk@classes.dex
data@app@com.paypal.paypahhere-2.apk@classes.dex
```

For example, to dump `data@app@com.example.example1-1.apk@classes.dex` into a file called `dump`, the command to be used is:

```
dexdump -d -f -h data@app@com.example.example1-1.apk@classes .dex > dump
```

The following is a screenshot of the kind of data collected in the dump file:

This dump will be in the form of the jump statements and is hard to read. A DEX de-compiler such as `baksmali` or `dedexer` can be used to make these files much more readable.

Summary

Security testing is a relatively young field. Patterns and testing strategies are still developing and security is being recognized as an important benchmark for identifying app weaknesses and improving app quality. In this chapter, we have put together our learning from all the previous chapters and used it to define test cases for our application. This is only a start and you should define test cases, as you deem appropriate for your use case.

We began with an overview of testing fundamentals. Then, we discussed the six pillars of security around which we designed our test cases. Some sample test cases were discussed that should provide you with a base for testing your application. And then we closed the chapter with a discussion of resources and tools that you could use for security-focused testing.

Let us now march to the last chapter of the book and see what is new and happening in the Android space that challenges our security basics.

10
Looking into the Future

You have made it to the last chapter, congratulations! So let's have some fun in this chapter and try to predict the future.

Mobile is a relatively new domain. It is in the phase of experimentation, where some technologies and use cases are successful while others may not get as much traction as expected. The focus of this chapter is to look at some technologies and use cases that are new to the mobile domain.

The chapter is divided into sections where each section discusses some experimentation with that technology or use case in mobile. We will begin with a discussion of mobile commerce with a focus on the product discovery, payments, and point of sale using mobile devices. Proximity technologies, such as NFC, RFID, and Bluetooth are discussed next. The following sections will talk about the use of mobile in health care and authentication. In the last section, we will discuss recent advancements in hardware from a security perspective.

Mobile commerce

Consumer behavior is changing commerce. These days commerce is not just a simple act of going to a merchant or a shop, selecting a product and paying for it. As illustrated in the following figure, with the emergence of new technologies mobile commerce includes product discovery by using geo-fencing, in-store and online research, payments by using self-scanning and self-checkout, sharing your purchases with your friends and then to managing your account. We also see a blurring line between online and offline commerce.

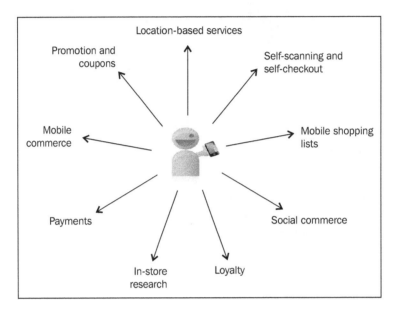

In the next few sections, we will discuss different components of commerce from a security perspective.

Product discovery using a mobile device

Product discovery is the process of finding a product. Merchants use different mechanisms to either bring customers to their retail stores or to encourage them to buy online. Product discovery also includes capabilities such as shopping lists, comparison shopping, and information about the product that facilitates a consumer to buy a product. A mobile device is ideal for this use case as a consumer can access information about a product and check the availability of a product in real time.

Some example application in the mobile space include applications for barcode scanning, location-based shopping, targeted advertising, points and perks to a user as they enter a retail store, ability to create shopping lists and be reminded of it when a consumer is close to a store that holds an item from the shopping list, and the ability to store loyalty cards in a wallet.

The biggest challenge from a security perspective is that of privacy. Targeted advertising and geo-fencing is based on the analysis of user data and their shopping patterns. Application developers should be aware of laws and regulations when using and collecting user data and preferences, and then using or sharing it. In almost all cases user consent is required before collecting information. This consent statement should include what is being collected and if it will be shared with third parties. Be mindful of updating this user consent as you add new functionality or update or extend existing functionality.

Mobile payments

Payment is the biggest component of mobile commerce. In any payment use case, there are three major entities: the consumer also called the buyer, the seller or the merchant and the infrastructure layer that enables payments.

Configurations

A consumer could be using a mobile device to search and pay for a product, a merchant could be using a mobile device, or both consumer and merchant could be using a mobile device. Ideally, all three entities are connected during an interaction. This is the case of full connectivity and is by far the most secure channel for payments. A user buying an item from eBay's mobile website is an example of full connectivity, as shown in the following figure:

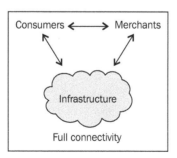

However, there are cases where they could be disconnected. A case where the consumer and merchant are both connected to the infrastructure but are not connected to each other is a case of infrastructure-centric connectivity. An example is geo-fencing when a user gets coupons for a store when they are close to it. In this case, the store and a user are both talking to the infrastructure (a third party or the carrier) but are not talking to each other. Another case is when a user a checks out with a device by using a point of sale terminal. In this case, a user uses the device as an authentication mechanism but may not be connected to the infrastructure layer. This is a case of merchant-centric connectivity in which a merchant is connected to both, the consumer and infrastructure but consumer is disconnected. Yet another example case is that of a consumer that talks to both the infrastructure and merchant but the merchant is disconnected. An example is when a user buys soda from a vending machine. The vending machine may sync with the backend at certain intervals of time and may be disconnected otherwise. The following figure illustrates partial connectivity configurations:

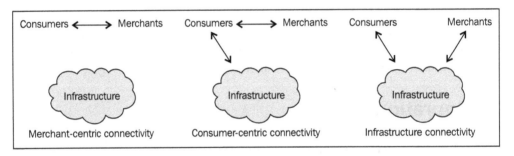

The basic security challenge in partial connectivity is that of end-to-end security. Since at any time there are two of the three connections, any stale state on client or server side is hard to detect. Then, there are issues with the client-merchant authentication, communication authentication, and privacy, as shown in the following figure:

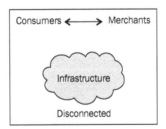

Then, there is the disconnected case where merchant and consumer talk to each other but none of them talks to the infrastructure layer. Maintaining the integrity of the device is a challenge in this case. Take the example of a consumer trying to use a coupon at the point of sale terminal.

The consumer may continue to use the coupon multiple times and the **PoS** (**Point of Sale**) terminal that cannot sync to the server to update the state of the coupon will not detect a fraud. Similarly, a client certificate may have expired or had been revoked but the merchant device will not be aware of it. If your application is set to work in such a scenario then there should only be a limited set of functionality available offline. Cases where PII or money is involved are best left to full connectivity or at least partial connectivity cases.

As an application, a developer should be aware of how your use case works. If your application can work with partial or no connectivity, you will need to take additional security measures when dealing with payments.

PCI Standard

Payment Card Industry (**PCI**) is an independent organization and works to create awareness about security in the payments use case. They have developed a common set of payment standards for ensuring user security is not compromised. PCI **PTS** (**Pin Transaction Security**) is for add-on devices that accept payments; PCI **P2PE** (**Point to Point Encryption**) is for hardware-based security, and PCI **DSS** (**Data Security Standard**) is for security management, policies, procedures, network architecture, software design, and other critical protective measures. The latest is version 2.0 and it helps organizations protect user data effectively. It has six core objectives that are implemented as the twelve core requirements. These are enumerated in the following figure:

Objectives	Requirements
Build and maintain a secure network	Install and maintain a firewall
	Do not use vendor-supplied defaults for system passwords and other security parameters
Protect Cardholder Data	Protect stored cardholder data
	Encrypt transmission of cardholder data across open, public networks
Maintain a Vulnerability Management Program	Use and regularly update anti-virus software on all systems commonly affected by malware
	Develop and maintain secure systems and applications
Implement Strong Access Control Measures	Restrict access to cardholder data by business need-to-know
	Assign a unique ID to each person with computer access
	Restrict physical access to cardholder data
Regularly Monitor and Test Networks	Track and monitor all access to network resources and cardholder data
	Regularly track systems and processes
Maintain an Information Security Policy	Maintain a policy that addresses information security

As an application developer working with payments, be aware of DSS. Payments are tricky and getting them right in a secure manner is a challenge in itself. So, you may like to use the already existing payment providers, such as PayPal.

More information about PCI can be found on their website available at `pcisecuritystandards.org`.

Point of Sale

Mobile Point of Sale (PoS) is a use that has been made possible with the ubiquity of a mobile device and by using proximity technologies as discussed earlier in this chapter. Your mobile device essentially acts as a Point of Sale terminal and can manage your ledgers and all your transactions for the day. Solutions provided by companies such as PayPal and Square use the phone audio jack to plug in a card swiping device. This device then reads the credit card details and in an encrypted form sends it over to the device. Other solutions include mobile Point of Sale terminals.

As an application developer, it is best to integrate with existing solutions instead of trying to invent the wheel. But, remember to ask some questions before you pick the solution. First, you need to ask if the solution provider is taking a proper security measure to encrypt the data. Please be aware of PCI DSS and PCI PTN, as we discussed in the previous sections. Retailers that process, store, or transmit credit card numbers must be PCI DSS compliant, or they risk losing the ability to process the credit card payments. Since there are infrastructural differences between credit cards in different countries, different technologies have to be employed to read credit/debit cards. In Europe, for example, the chip and pin technology is the norm, so your PoS payment provider should have solutions in each area. You might like to choose a provider with whom you can manage your credit cards and also checks, cash and other form of payments.

The previous figure shows some examples of mobile Point of Sale solutions. The first image is of a PayPal card reader in North America and the application that manages all modes of payments.

The preceding figure is PayPal's pin and chip solution in Europe that works by using Bluetooth.

Above is another example of mobile point of sale. Most commonly used by delivery agents and sales representatives.

Proximity technologies

Proximity technologies work in a radius of inches or centimeters. These include technologies such as **Near Field Communication** (**NFC**), Bluetooth, and **Radio Frequency Identification** (**RFID**). Most of these technologies have been around for some time but the pervasiveness of mobile devices has given them an array of new use cases. These technologies are now being used for mobile payments, pairing of different devices, identification, and authentication.

Bluetooth is now a standard in most mobile phones. It is a wonderful technology to pair devices. With devices such as glasses and watches coming in the market, this might be the technology that brings them all together.

Both NFC and RFID work by producing an electromagnetic field that modulates at a certain frequency. Since these tags are world readable, when used as tags or identification mechanism, these tags pose privacy risk. The first NFC enabled Android phone, the Nexus S, came out in the year 2010. The Android SDK comes bundled with the API for using the NFC tags.

Owing to their small range of operation, proximity technologies are mistakenly considered to be secure. However, it is not the case. A quick search will reveal exploit scenarios for all cases. Data modulation, jamming of data, and privacy are some risks associated with these technologies.

Social networking

An array of social networking applications exist in the App Stores today and the new use cases are being tested each day. These applications let friends, acquaintances, neighbors, colleagues, and people with special interests share, collaborate, and essentially keep in touch with each other. Some successful examples include Facebook, Twitter, Pinterest, Google Hangout, and LinkedIn.

Social networks work as graphs of networks linking entities together. Any bad node in the graph has the potential to spam or infect the other node. In the following figure, the message between nodes A and B is intercepted and changed with spam. This will result in all nodes connected to B to be infected. This continues and as you can imagine will spread across nodes very quickly:

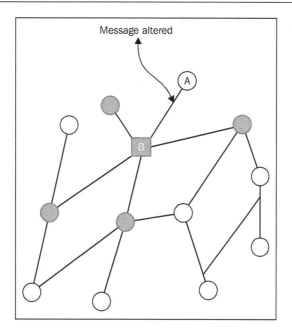

The biggest challenge with social networking applications is that of privacy. First, users have to be mindful of what they share with their contacts. In most cases, users are using their real name and other private information.

Second, users need to be aware of spam and malware. Not everyone is your friend. Not all games that your friends played with are written by good guys. There is also no need to click on all the links shared by people you are following.

Third, application developers have to be mindful of how they store and process users sensitive information. The first line of defense is to specifically ask a user what they want to share and with whom. This user consent should save the developers from liability issues. Second, they have to define proper access control based on user preferences. Third, they have to secure user details and PII both at rest and in transit.

Another issue with social networking sites is that of identity theft. It is easy for a malicious user to create an account by using other person's identity.

Healthcare

Developing mobile applications for healthcare is another example of a very security sensitive use case. In the healthcare use case a developer is dealing with user identification, electronic medical histories, laboratory tests, and prescribed medicines. Compromising this information may affect the health of a patient.

Mobile devices can be put to great use in healthcare as they are very personal and we carry them with us always. So, applications those remind us to take our pills on time, doctor visit, notes taking applications for both doctors and patients, instant notification about laboratory results, and reminders that are prescription medication needs to be refilled are all important and useful applications.

Mobile devices can also be used in emergency situations where other people can help an ailing person with the aid of a mobile device. Users can share real time video and talk to a doctor in real time to get help.

The other arm of development in healthcare is the use of the Android platform in embedded devices, such as scanners, radiology, X-ray machines, robotic surgery, and ultrasound devices.

Accurate identification of a person is crucial in healthcare. Also remember the important security rule: trust but verify. So, you identify a person but want to do it one more time to be sure. Access control and secure storage and transmission of PII are important as well.

Be aware of standards and regulations in the healthcare field such as **Health Insurance Portability and Accountability Act (HIPAA)**.

Authentication

Authentication is the act of identifying an entity. In our case, authentication usually relates to identifying a person. The current method of authentication is by using a username and password. Since, passwords are complicated and are hard to type on a small device, phone number and PIN are used for authenticating a user.

Two-factor authentication

The most common approach today is two-factor authentication. This is based on the theory that to identify a person uniquely, a person should provide two of the following three identifiers:

- Something that a user has; this includes a digital signature, security tokens, phone, tag, and so on

- Something that a user knows; this includes passwords, secrets, PIN, or an answer to a question that only the user is expected to know

- Something that a user is; examples include retina scan, fingerprints, and facial recognition

An example of two-factor authentication is logging in by using username/password or phone/PIN followed by entering a secret code sent in an SMS to the user device. Another example could be entering a username and password and then answering a challenge question, as shown in the following image:

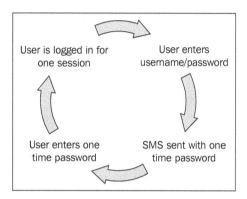

Implementing a two-factor authentication is easy in Android. Google authenticator implements the two-factor authentication by using SMS or voice call.

Biometrics

Biometric authentication is an identification of a user by using biological properties that are unique to a user. These include the use of fingerprints, facial recognition, retina scan, and iris scan. Based on the iris scan, India has implemented the world's largest identification system called Aadhar. This ambitious project will provide a unique number to all Indian citizens aged five and above by using their demographic and biometric information. Check out the website of **Unique Identification Authority of India (UIDAI)** available at www.uidai.gov.in.

There are some applications on Android which use biometrics as key. The important consideration when using such an application is to be sure that the user identification specifications are not stored on the device. Second, if this information is stored on the server, how is it transmitted and stored there? Third, how do you access this information?

Biometrics is unlike the two-factor authentication where you can easily change your password or update your RSA secure ID token. Biometrics is personal, and compromising this information comes with huge risks.

There are two scenarios in which biometrics is used. In the first case, a user is verified by using some biological attribute, such as a fingerprint. This is compared to the copy stored in the device. Logging into a phone by using your face is an example of verification, as follows:

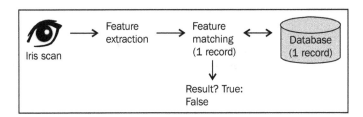

The second case is that of identification in which a biometric identity is compared against stored identities in a database for a match. The biometric identification system being implemented in India is such an example. The following figure illustrates this process:

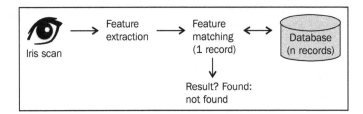

Advances in hardware

Mobile operating systems have come a long way. When I started working on mobiles, we had candy bar phones that did little more than make phone calls and had basic utilities such as calculator and a widget to show the time and date. To support the advanced use cases for mobile, security has to be built in the hardware itself. There are some efforts in this direction that I have discussed in the following sections.

Hardware security module

Hardware security module, also called a secure element, is a piece of hardware (chip) embedded in the hardware to store cryptographic keys and other sensitive information. The idea is to provide an isolated, tamper resistant environment to store PII. In some cases, a secure element can be carried with the device as well. Examples of secure elements include an enhanced SIM card controlled by a mobile network operator, a chip embedded in the device itself, or a micro SD card with a special circuit built-in. Many Android phones come equipped with a secure element.

In some cases, security modules also work as security accelerators. These accelerators, besides storing keys, also perform crypto functions in the hardware, such as encryption, decryption, hashing, and random number generation. This offers a huge offload from the CPU and results in better performance.

For developers to be able to use a secure element, it has to be exposed through APIs. Android's **Secure Element Evaluation Kit (SEEK)** for Android is a step in this direction. Based on open mobile APIs, the aim of this set of APIs, called Smart Card APIs, is to provide a mechanism for the application to communicate with embedded secure elements, SIM cards, or other device encryption modules. Check out more information available at `code.google.com/p/seek-for-android`. The following image from `code.google.com` illustrates the concept of SEEK very efficiently:

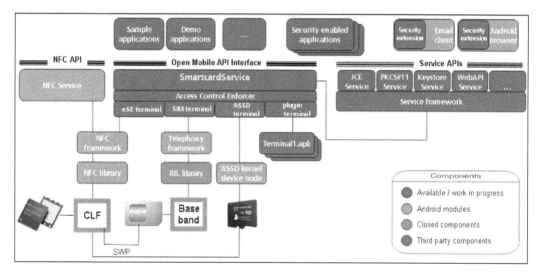

Based on Android's permission mechanism, Smart Card APIs need a special permission called `android.permission.SMARTCARD` for applications accessing these APIs. The Smart Card API remote process is registered with a unique UID/GID of the Smart Card. Note that this security mechanism ceases to work on rooted devices. `GoogleOtpAuthenticator` is implemented over a Smart Card API using two-factor authentication.

TrustZone

Developed by ARM, and now with GlobalPlatforms, TrustZone technology is complete security solution for devices. It is based on systems-on-chip, TrustZone provides a trusted execution environment for applications such as payments, content streaming and management, access control, and other PII. The cool feature of TrustZone is that each application runs in its own contained environment completely isolated from each other.

You may like to check out the website `www.arm.com/products/processors/technologies/trustzone.php` for details. The following figure from the preceding website shows a high-level view of this technology. Many mobile processors such as from Texas instruments and Nvidia's Tegra core are built upon the TrustZone technology.

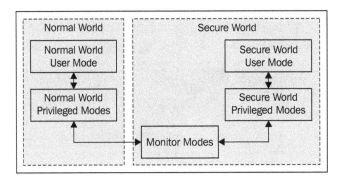

As shown in the previous figure, using virtualization, the processor is divided into two virtual domains: one for the normal mode and the other for executing sensitive processes called the secure mode. By using a monitor mode, the process transitions from one mode to the other. All sensitive code, data, and resources are processed away from the normal operating environment, software, and memory on the device. This isolation is enforced by SoC architecture so it is highly robust against software and probing attacks.

Mobile trusted module

In 2010, **Trusted Computing Group** (**TCG**) published the 1.0 version of **Mobile Trusted Module** (**MTM**). TCG is an international standards body that works with its members to develop standards and specifications. MTM's aim is to adapt an existing TCG technology for mobile and embedded use.

Trusted computing is based on a hardware root of trust and is called the **Trusted Platform Module** (**TPM**). It detects malware and checks the integrity of a system. This capability is called the Trusted Platform Module. The security of TPM starts with the boot process. A hardware root of trust (usually a key) is burned in the processor itself. Boot security is built on this root of trust. Progressive stages of the boot software are verified cryptographically to ensure that only correct, authorized software is executed in the device.

Check out their website available at `www.trustedcomputinggroup.org`. It is more relevant for kernel developers but it makes for a very interesting read for anyone.

Application architecture

These days there are three ways to write an application: native, mobile web, and hybrid.

A native application is specific to a platform and is written in a language that is native to the platform. These applications use native tools and SDKs that are provided by the operating system manufacturer. These applications have much better performance and can use native features and APIs for secure data storage. The following figure illustrates how native and hybrid applications work:

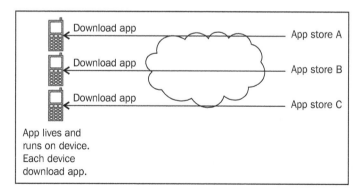

A mobile web application is written with web technologies, such as HTML5, CSS, PHP, JavaScript, and ASP.net. These applications are cross-platform and once they are written they can be run on any platform that has a browser. They provide the ease of centralized updates but inherit all the browser vulnerabilities. Be aware of the browser exploits when you write a mobile web application. Browser code is easily available for everyone to see. Also, URL exploits are a risk in such applications, as the application does not reside on the device and can be accessed only by using a valid URL. The following is a figure illustrating how mobile web application works:

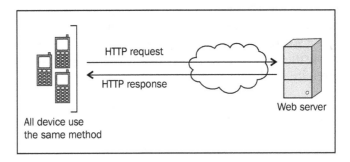

The third way to write an application is to develop a hybrid application. This application combines the benefits of both, native and mobile web. An application is written once by using web technologies. The user needs to install the application just like a native application and it runs in a native browser by using the device's browser engine. In this way the application can run in the offline mode, can access device capabilities, and a developer can target multiple platforms.

The decision to pick which architecture to use rests on your use case. Native applications are much more secure than hybrid or mobile web. They also perform better in terms of speed and user experience. Hybrid and mobile web application, on the other hand, are easier and quicker to develop by using web technologies and are cross-platform.

Summary

This chapter focused on the upcoming use cases and technologies and how they relate to mobile security in general. We discussed mobile commerce, proximity technologies, mobile security in healthcare, and authentication. We closed the chapter with a look at the security enhancements in the hardware space. As you will have noticed, there is a lot happening in the mobile space and I think it will continue this way for a while before things settle down.

With this we have reached the end of this book. I hope you learnt something new in this book and enjoyed this journey as much as I did.

Index

U

UID 10
UIDAI
  about 181
  URL 181
unbindService() 26
Unique Identification Authority of India.
        See UIDAI
unit testing 156
Unknown Sources option 150
URI (Universal Resource Identifier) 55
usability
  versus security 164
usability tests 158
user data
  storing 86, 87
User Identification. *See* UID
user preferences
  preference activity 125
  shared preferences 123

uTest
  URL 161

V

Virtual Private Network. *See* VPN
VPN 148

W

WAP (Wireless Application Protocol) 89
WiMAX (Worldwide Interoperability for
        Microwave Access) 89

X

XORed (Exclusive OR) 109

Z

Zygote 11

Thank you for buying
Android Application Security Essentials

About Packt Publishing

Packt, pronounced 'packed', published its first book "*Mastering phpMyAdmin for Effective MySQL Management*" in April 2004 and subsequently continued to specialize in publishing highly focused books on specific technologies and solutions.

Our books and publications share the experiences of your fellow IT professionals in adapting and customizing today's systems, applications, and frameworks. Our solution based books give you the knowledge and power to customize the software and technologies you're using to get the job done. Packt books are more specific and less general than the IT books you have seen in the past. Our unique business model allows us to bring you more focused information, giving you more of what you need to know, and less of what you don't.

Packt is a modern, yet unique publishing company, which focuses on producing quality, cutting-edge books for communities of developers, administrators, and newbies alike. For more information, please visit our website: www.packtpub.com.

Writing for Packt

We welcome all inquiries from people who are interested in authoring. Book proposals should be sent to author@packtpub.com. If your book idea is still at an early stage and you would like to discuss it first before writing a formal book proposal, contact us; one of our commissioning editors will get in touch with you.

We're not just looking for published authors; if you have strong technical skills but no writing experience, our experienced editors can help you develop a writing career, or simply get some additional reward for your expertise.

Android 3.0 Application Development Cookbook

ISBN: 978-1-84951-294-7   Paperback: 272 pages

Over 70 working recipes covering every aspect of Android development

1. Written for Android 3.0 but also applicable to lower versions

2. Quickly develop applications that take advantage of the very latest mobile technologies, including web apps, sensors, and touch screens

3. Part of Packt's Cookbook series: Discover tips and tricks for varied and imaginative uses of the latest Android features

Android 4: New Features for Application Development

ISBN: 978-1-84951-952-6   Paperback: 166 pages

Develop Android applications using the new features of Android Ice Cream Sandwich

1. Learn new APIs in Android 4

2. Get familiar with the best practices in developing Android applications

3. Step-by-step approach with clearly explained sample codes

Please check **www.PacktPub.com** for information on our titles

Android Application Testing Guide

ISBN: 978-1-84951-350-0   Paperback: 332 pages

Build intensiely tested and bug free Android applications

1. The first and only book that focuses on testing Android applications

2. Step-by-step approach clearly explaining the most efficient testing methodologies

3. Real world examples with practical test cases that you can reuse

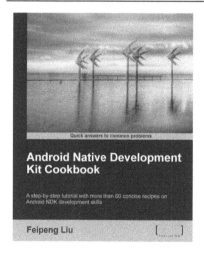

Android Native Development Kit Cookbook

ISBN: 978-1-84969-150-5   Paperback: 346 pages

A step-by-step tutorial with more than 60 concise recipes on Andriod NDK development skills

1. Build, debug, and profile Android NDK apps

2. Implement part of Android apps in native C/C++ code

3. Optimize code performance in assembly with Android NDK

www.ingramcontent.com/pod-product-compliance
Lightning Source LLC
Chambersburg PA
CBHW080408060326
40689CB00019B/4172

* 9 7 8 1 8 4 9 5 1 5 6 0 3 *